Getting the best from t

We have designed this book to give you all the support you nee̶ ... ̶ ̶ary for success on your course. The science, maths and quality of w̶ ̶ ̶ ̶ ̶ ̶ ̶ ̶ ̶ ̶ ̶ ̶ ̶ ̶ ̶ ̶cation skills (QWC), for every major exam specification, are explained in detail.

There are activities to practise each and every skill so you have a chance to apply your learning. They are set in interesting contexts and give you the chance to use your skills in unfamiliar situations. We have included the answers so you can check your understanding and also see how you can improve using the helpful hints, tips and pointers. The Assessing Investigative Skills activities give you the opportunity to practise some of the skills you will require in assessed practical tasks. To help you improve the quality of your written communication, we have included worked examples to show how low, medium and high answers get their marks. There is also a glossary to check your understanding of key terms.

SKILLS

Each skill is colour-coded to tell you what type of skill it is:
Working Scientifically
Quality of Written Communication
Maths.

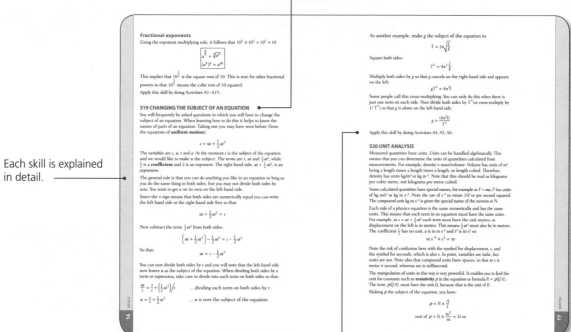

Each skill is explained in detail.

The activities that practise the skill are listed at the bottom.

ACTIVITIES

The activities are listed in a logical sequence to allow you to tackle increasingly complex ideas as your knowledge of physics deepens.

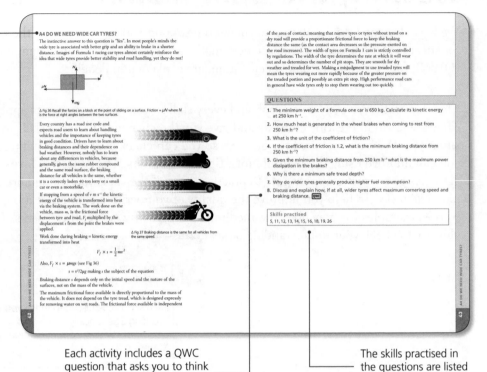

A4 DO WE NEED WIDE CAR TYRES?

The instinctive answer to this question is "Yes". In most people's minds the wide tyre is associated with better grip and an ability to brake in a shorter distance. Images of Formula 1 racing car tyres almost certainly reinforce the idea that wide tyres provide better stability and road handling, yet they do not!

△ Fig 36 Recall the forces on a block at the point of sliding on a surface. Friction = μN where N is the force at right angles between the two surfaces.

Every country has a road use code and expects road users to learn about handling vehicles and the importance of keeping tyres in good condition. Drivers have to learn about braking distances and their dependence on bad weather. However, nobody has to learn about any differences in vehicles, because generally, given the same rubber compound and the same road surface, the braking distance for all vehicles is the same, whether it is a correctly laden 40-ton lorry or a small car or even a motorbike.

If stopping from a speed of v m s^{-1} the kinetic energy of the vehicle is transformed into heat via the braking system. The work done on the vehicle, mass m, is the frictional force between tyre and road, F_f multiplied by the displacement s from the point the brakes were applied.

Work done during braking = kinetic energy transformed into heat

$$F_f \times s = \tfrac{1}{2}mv^2$$

Also, $F_f \times s = \mu mgs$ (see Fig 36)

$$s = v^2/2\mu g \text{ making } s \text{ the subject of the equation}$$

Braking distance s depends only on the initial speed and the nature of the surfaces, not on the mass of the vehicle.

The maximum frictional force available is directly proportional to the mass of the vehicle. It does not depend on the tyre tread, which is designed expressly for removing water on wet roads. The frictional force available is independent

△ Fig 37 Braking distance is the same for all vehicles from the same speed.

of the area of contact, meaning that narrow tyres or tyres without tread on a dry road will provide a proportionate frictional force to keep the braking distance the same (as the contact area decreases so the pressure exerted on the road increases). The width of tyres on Formula 1 cars is strictly controlled by regulations. The width of the tyre determines the rate at which it will wear out and so determines the number of pit stops. They are smooth for dry weather and treaded for wet. Making a misjudgment to use treaded tyres will mean the tyres wearing out more rapidly because of the greater pressure on the treaded portion and possibly an extra pit stop. High performance road cars in general have wide tyres only to stop them wearing out too quickly.

QUESTIONS

1. The minimum weight of a formula one car is 650 kg. Calculate its kinetic energy at 250 km h^{-1}.
2. How much heat is generated in the wheel brakes when coming to rest from 250 km h^{-1}?
3. What is the unit of the coefficient of friction?
4. If the coefficient of friction is 1.2, what is the minimum braking distance from 250 km h^{-1}?
5. Given the minimum braking distance from 250 km h^{-1} what is the maximum power dissipation in the brakes?
6. Why is there a minimum safe tread depth?
7. Why do wider tyres generally produce higher fuel consumption?
8. Discuss and explain how, if at all, wider tyres affect maximum cornering speed and braking distance. **QWC**

Skills practised
5, 11, 12, 13, 14, 15, 16, 18, 19, 26

Each activity includes a QWC question that asks you to think about your quality of written communication.

The skills practised in the questions are listed at the bottom of each activity.

There are two Assessing Investigative Skills activities.

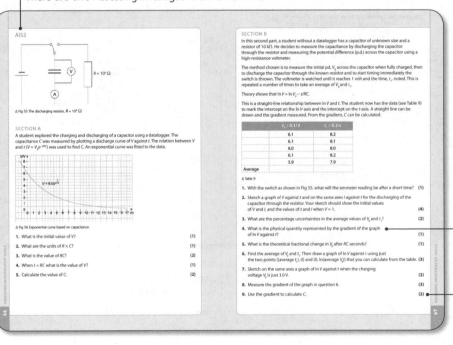

AIS2

△ Fig 55 The discharging resistor, $R = 10^4\,\Omega$

SECTION A

A student explored the charging and discharging of a capacitor using a datalogger. The capacitance C was measured by plotting a discharge curve of V against t. The relation between V and t ($V = V_0 e^{-t/RC}$) was used to find C. An exponential curve was fitted to the data.

$$V = 8.0 e^{-0.5t}$$

△ Fig 56 Exponential curve based on capacitance.

1. What is the initial value of V? (1)
2. What are the units of $R \times C$? (1)
3. What is the value of RC? (2)
4. When $t = RC$ what is the value of V? (1)
5. Calculate the value of C. (2)

SECTION B

In this second part, a student without a datalogger has a capacitor of unknown size and a resistor of 10 kΩ. He decides to measure the capacitance by discharging the capacitor through the resistor and measuring the potential difference (p.d.) across the capacitor using a high-resistance voltmeter.

The method chosen is to measure the initial p.d. V_0 across the capacitor when fully charged, then to discharge the capacitor through the known resistor and to start timing immediately the switch is thrown. The voltmeter is watched until it reaches 1 volt and the time, t_1, noted. This is repeated a number of times to take an average of V_0 and t_1.

Theory shows that $\ln V = \ln V_0 - t/RC$.

This is a straight-line relationship between $\ln V$ and t. The student now has the data (Table 9) to mark the intercept on the ln V-axis and the intercept on the t-axis. A straight line can be drawn and the gradient measured. From the gradient, C can be calculated.

$V_0 \pm 0.1$/V	$t_1 \pm 0.2$/s
6.1	8.2
6.1	8.1
6.0	8.0
6.1	8.2
5.9	7.9
Average	

△ Table 9

1. With the switch as shown in Fig 55, what will the ammeter reading be after a short time? (1)
2. Sketch a graph of V against t and on the same axes I against t for the discharging of the capacitor through the resistor. Your sketch should show the initial values of V and I, and the values of t and I when $V = 1$. (4)
3. What are the percentage uncertainties in the average values of V_0 and t_1? (2)
4. What is the physical quantity represented by the gradient of the graph of ln V against t? (1)
5. What is the theoretical fractional change in V_0 after RC seconds? (1)
6. Find the average of V_0 and t_1. Then draw a graph of ln V against t using just the two points ((average t_1), 0) and (0, ln(average V_0)) that you can calculate from the table. (3)
7. Sketch on the same axes a graph of ln V against t when the charging voltage V_0 is just 3.0 V. (2)
8. Measure the gradient of the graph in question 6. (3)
9. Use the gradient to calculate C. (2)

The questions help you to practise some skills in a practical context. They are similar to the practical assessments you will encounter on your course.

Each question includes an indication of the number of points needed in your answer.

The answers include helpful tips and hints, often with fully worked out examples to help you understand what you need to do.

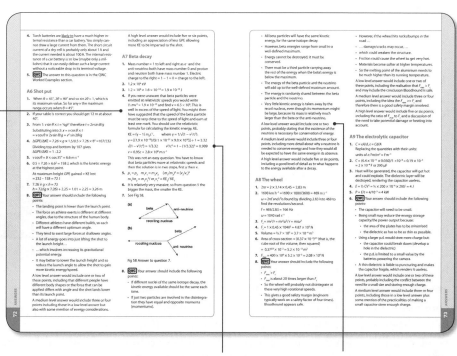

Fully worked-out approaches are included to help you build up the skills you require.

Each QWC answer includes low, medium and high level answer pointers to help you improve.

The QWC Worked Examples show low, medium and high written responses.

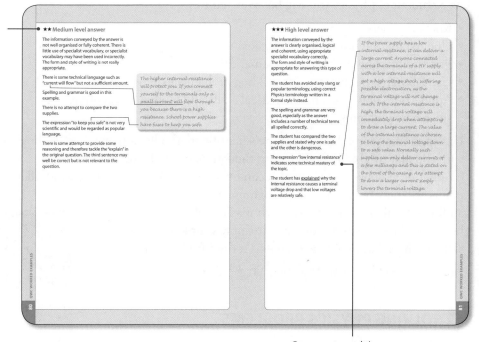

Comments explain how to improve the quality of written communication.

GETTING THE BEST FROM THE BOOK

Skills

S1 FROM HYPOTHESIS TO THEORY

The sum total of physics knowledge is physics theory. The word *theory* is synonymous with the word *knowledge*. Physicists often have ideas, but these cannot become theory until they have been tested experimentally.

An hypothesis is a proposal for testing an idea using current theory. Newton was famously asked what the mechanism was for action at a distance in his new Law of Gravity. He responded "hypothesis non fingo", which means that he was unable to come up with a hypothesis and so would not speculate with any untestable ideas.

△ Fig 1 Isaac Newton 1643–1727.

If an idea is supported by an experimental test then the test will be repeated. If the results are similar to the earlier ones, they are considered **repeatable** and the idea will be published and communicated to other scientists. They in turn will test the idea, either with similar experiments or new ones. If the test is similar and the idea is again supported then it is said to be **reproducible** and this provides further support for the idea. If the test is a new one and supports the idea, this is further evidence that corroborates the idea. In either case the idea can now take its place as part of physics theory. Other scientists may still contest the methods used if the idea is a controversial one; in this case the idea may have to wait for more supporting evidence.

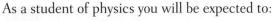

As a student of physics you will be expected to:

• Know the facts of physics.
• Understand the principles of physics.
• Practise tackling problems mathematically.
• Use laboratory techniques for tackling problems experimentally.
• Apply your knowledge, mathematical skills and laboratory techniques to solve new problems.

Understanding something about the scientific method will help when dealing with the first two bullet points, even though facts can be found on the internet or in books and principles can be learned. The third and fourth bullet points will only be developed through practice and this will improve your ability to think like a physicist. With mastery of the first four bullet points you will be equipped to apply your knowledge and skills to the solution of new problems – the fifth bullet point.

Apply this skill by doing Activities A1, A7, A10.

S2 SOLVING PHYSICS PROBLEMS

Solving physics problems means finding a method or process. This may be obvious from the way the question is asked, but if it is not clear how to proceed, you can adopt the following order: energy, momentum and then force (emf!).

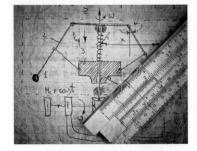

△ Fig 2 Modern calculators, rather than slide rules, are now part of our everyday life.

Energy

It is often easy to solve problems by considering energy conservation. Problems arise when some energy in the transformation cannot be accounted for – usually where energy is dissipated as heat. For example, energy dissipation during collisions means that when the collision is inelastic some of the original kinetic energy is dissipated as heat and this may be difficult to account for. While energy conservation applies in all situations, there are many ways of calculating energy and you may have insufficient data to do so. Energy, however, is the easiest way to begin.

Momentum

In all collisions, elastic or inelastic, momentum is conserved. This may help where the law of conservation of energy is difficult to apply. The only difficulty comes from making sure that you identify correctly the *participants* in the collision. There is only one way of calculating momentum, unlike energy. You may be able to analyse a problem by considering impulse and change in momentum.

Force

Analysing a problem by considering forces is usually the most difficult process. You often have to draw a free body diagram (see Skill 5) and, although this appears simple, mistakes are easily made. You probably began studying forces first, but unless you are specifically asked to carry out a force analysis you should keep this until last.

Apply this skill by doing Activity A3, A10.

S3 ENERGY CONSERVATION

Julius Mayer and Hermann von Helmholtz are credited with formulating the law of conservation of energy or the first law of thermodynamics in the early 1840s. Mayer published first and even calculated the mechanical equivalent of heat before James Joule.

△ Fig 3 Does it add up to a sustainable future?

The law: in all processes energy is conserved; that is, the amount of energy at the beginning of the process is equal to the amount of energy at the end of the process. It is simply transformed or transferred between systems.

Over the last century and a half, methods of calculating energy in its various forms have been devised. The important fact is that all these methods are internally consistent *if* the correct units are chosen. The problem physicists face is that energy is dissipated during most practical processes. That is, some energy almost always ends up as heat and may be difficult to quantify. When trying to understand exactly what energy is, you are unlikely to be satisfied with any answer.

Whenever work is done, energy is transferred by mechanical processes.

Work, then, is energy and is measured in joules (symbol J). The second law of thermodynamics can be interpreted as saying that some work always ends up as dissipated heat so that perpetual motion is impossible. An allied concept is entropy, which is a measure of disorder. As some work is always dissipated, entropy is always increasing in the Universe.

From Einstein's equation $E = mc^2$, it is clear that energy and mass are related only by a constant, c, and not by any other variable, so we say that they are equivalent. Mass can, in principle, be measured in joules and energy could be measured in kilograms. The concept is hard to grasp but effectively it means that whenever there is an energy change (or work is done) there is a corresponding change in mass. This is the principle of conservation of mass-energy.

Apply this skill by doing Activities A6, A10, A11.

S4 MOMENTUM CONSERVATION

The law of conservation of momentum, p, is derived from Newton's second and third laws. Momentum is a **vector** quantity and is always conserved, whether collisions are elastic or inelastic. The momentum before a collision equals the momentum after the collision. In totally *inelastic collisions*, the bodies move together or remain stationary after the collision. The key to understanding what is going on is that different amounts of kinetic energy (KE) are being dissipated. For totally *elastic collisions*, the KE before and after the collision is conserved as well as the momentum. In other collisions some of the KE is transformed into heat. If the initial momenta are equal and opposite, all the KE may be transformed into heat.

Δ Fig 4 Elastic collisions in Newton's Cradle.

When the interaction is between an object of small mass and a relatively large mass object, the large object will hardly move so that very little kinetic energy is transferred to the large object. If one of the objects is relatively large and *stationary*, its momentum will be close to zero after the collision, as it was before.

Apply this skill by doing Activity A7.

S5 FREE BODY DIAGRAMS

Using arrows to represent forces is a way of modelling physics problems. Forces arise out of interactions between two or more bodies and can be gravitational, electrical, magnetic or nuclear. When considering the forces on a single body, you must isolate it from its surroundings and be careful to draw all the forces that act on it. This means not confusing forces that act on other bodies. You need to be careful to ensure that the forces you identify are in the correct direction and position, and that you have identified them all. Use the process below when carrying out a force analysis.

1. First isolate your body.

2. Then draw all the forces acting on it.

3. Then resolve the forces into two sensible perpendicular directions.

4. Then sum the forces in each direction.

5. Finally, find the **resultant** force using Pythagoras' theorem and trigonometry.

The free body diagram in Fig 5 shows the forces on a uniform ladder leaning against a rough wall and resting on rough ground. It shows the forces at stage 3 above, where the forces on the top and bottom of the ladder have been resolved into two sensible perpendicular directions. As the ladder is uniform the weight of the ladder acts from the middle. The ladder is in equilibrium so that the resultant force in each perpendicular direction is zero.

Apply this skill by doing Activities A4, A14.

S6 FLUX AND FIELD LINES

We owe the flux and lines of force model for fields to Michael Faraday. He needed a visual model for magnetic fields as he was unable to develop a mathematical model due to the complex shapes of most magnetic fields. Gravitational and electric fields can also be modelled using field lines. Gravitational and electric field lines begin and end on masses and charges, respectively.

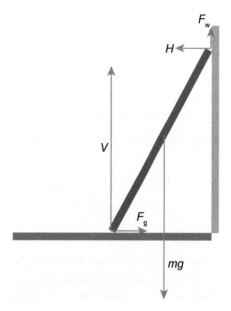

△ Fig 5 Forces on a ladder.

Magnetic field lines begin at magnetic north and end on magnetic south poles but also form closed loops around current-carrying conductors. The existence of an electric current can be deduced from the presence of a magnetic field even if it is the tiny atomic currents present in the material of permanent magnets.

Masses and charges, whether or not they are already moving, experience forces along gravitational and electric field lines, respectively. Charges that move at an angle to magnetic field lines experience magnetic force perpendicular to magnetic field lines. Force fields are therefore **vector** in nature – at each point of the field an object will experience a force in a particular direction. The concept of field provides one of the most important models in physics.

Magnetic flux, denoted as Φ, simply relates to the number of field lines, whereas magnetic field strength, B, is how closely the field lines are packed together. The unit of flux Φ is the weber, Wb.

△ Fig 6 Charged particles guided by magnetic field lines on the Sun's surface are revealed in giant prominences.

Field strength is also referred to as flux density so that $\Phi = B \times A$ webers, where A is the area perpendicular to the field lines through which the flux acts. Of course, nobody actually knows how many field lines there are because they are imaginary, but formulae have been constructed that allow you to calculate B and so calculate Φ.

The field strength inside a long solenoid is proportional to the current through the windings. As the cross section of the solenoid increases, the field will remain the same if the current does not change. This means the number of flux lines must increase to keep B the same.

You will need the concept of flux density or field strength, B, to calculate forces on current carrying conductors, For example, $F = BIl$ where l is the length of the conductor carrying a current.

You will need the concept of flux to calculate induced potential difference (or emf or voltage), for example $V = \Delta\Phi / \Delta t$, which says that the induced potential difference is proportional to the rate of change of flux across a conductor.

Apply this skill by doing Activities A12, A15.

S7 UNCERTAINTY (ABSOLUTE)

Uncertainty and *error* are frequently used interchangeably. Error should be reserved for the difference in the measured or calculated quantity from the generally accepted value. For most measurements there are no accepted values. For example, you measure your height and find 1.72 m. There is an uncertainty in this measurement based on the method of measurement. The ruler may be calibrated in mm but your measurement implies an uncertainty of 1 cm so is written 1.72 ± 0.01 m. The value 0.01 m is the absolute uncertainty in the measurement.

However, you kept your shoes on so you will need to subtract the thickness of the heel. This is measured as 3 ± 1 cm or 0.03 ± 0.01 m. Clearly, the uncertainty is compounded when the measurements are subtracted and the absolute uncertainties are added, giving 1.69 ± 0.02 m.

In general, when adding or subtracting measurements the absolute uncertainties are added.

Another measurement is made of the heel thickness, 3.0 ± 0.01 cm or 0.030 ± 0.001 m. The **precision** of the heel measurement is better than that of the height measurement. In this case, when the measurements are added or subtracted the answer is only as precise as the least precise measurement. There is no uncertainty in the first figure of the heel measurement so the result is 1.69 ± 0.01 m.

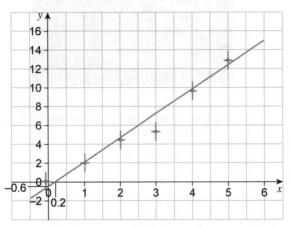

△ Fig 7 Representing uncertainty when graphing measurements.

When graphing measurements the uncertainty can be represented by extending the point vertically and horizontally as well if required to create "error bars". The best fit line is then expected to fall within these error bars. In this way the maximum and minimum value for the gradient can be determined. The uncertainty in the gradient is then half of the difference between maximum and minimum.

Apply this skill by doing Activities AIS1, AIS2.

S8 UNCERTAINTY (PERCENTAGE)

After measuring your height and finding 1.69 ± 0.01 m, you measure your body mass to determine your body mass index (BMI) and find 75.5 ± 0.5 kg.

$$BMI = mass/(height^2)$$

In this case a quantity, BMI, is being calculated by multiplying and dividing measurements, and the fractional errors in each measurement must be added. The easiest way to do this is to add the percentage errors in the measurements.

For the BMI calculation the percentage errors are $[(0.01/1.69) \times 100]$ for the height and $[(0.5/75.5) \times 100]$ for the mass. These are approximately 0.6% and 0.7%, respectively (rounded up). Since the height is squared the total percentage error in BMI is 1.9%.

Calculation gives BMI = 26.43465 to seven figures. The total percentage error shows that the absolute error is $1.9\% \times 26.43465$, or 0.5 to one figure. It only makes sense to keep one or at most two figures in the error. This uncertainty then determines the last **significant figure** (see also S25) in your result, so that your BMI is

$$BMI = 26.4 \pm 0.5 \text{ kg m}^{-2}$$

When multiplying or dividing measurements add the percentage errors; then use the total percentage error to find the absolute error in the result.

A general rule is to only keep as many significant figures in the result as there are in the least significant measurement.

Apply this skill by doing Activities AIS1, AIS2.

S9 SYSTEMATIC AND RANDOM ERROR

Measuring instruments themselves may be in error, usually with **zero errors**, where the instrument does not read zero when the variable value is zero. These will result in **systematic errors** appearing in measurements and results. These errors often show up when best fit straight line graphs of direct proportionality do not go through the origin.

An experiment to measure the density of aluminium involves measuring the masses and volumes of different pieces using a measuring cylinder of water. The results are shown in Table 1.

The gradient of this graph (Fig 8) is 2.8 to 2 significant figures. Continuation of the line reveals that it does not go through the origin but cuts at about -0.7 g. It is possible that volume measurements were not made to the bottom of the meniscus. The observer has systematically read a greater volume than the true volume, resulting in an effective zero error of the measuring cylinder. Other zero errors, such as forgetting to **tare** the balance, so systematically reading a mass that is too small, would also give rise to a systematic error, resulting in a similar deviation from the origin.

Volume/cm³	Mass/g
1.5	3.6
2.1	4.8
2.4	6.2
2.9	7.6
3.7	9.7
4.6	12.2

△ Table 1 Experimental results.

Random errors made by the observer are more correctly referred to as uncertainty.

Using our very simple technique of taking the precision of the results of calculations only as far as the least precise measurement gives the density of aluminium as 2.8 g cm^{-3} with an uncertainty of 0.1g cm^{-3}. In fact the accepted value of the density of aluminium is 2.70 g cm^{-3} so that the error here is no bigger than the uncertainty.

In general if the error is bigger than the uncertainty we would start looking for systematic errors, which would increase the overall uncertainty in the result.

Apply this skill by doing Activities AIS1, AIS2.

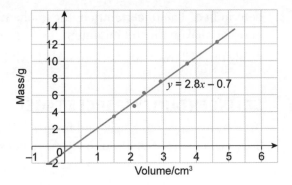

△ Fig 8 Graph of Table 1 experimental results.

S10 RISK ASSESSMENT

You may be asked to make some sort of risk assessment before carrying out practical work or evaluating a written account of a technique. This involves first identifying the hazards, such as acid hazard, that may be encountered. Risk of actual harm arises when considering the likelihood and severity of the hazard occurring.

Table 2 assigns values to likelihood down the side and severity along the top. The risk from the hazard can then be calculated from likelihood × severity, shown in the matrix in Table 2. In the example shown, acceptable risk might be considered anything up to 4. For risks greater than this you would be expected to describe what additional safety or control measures you would put in place. Risk assessment is a subjective process, its principal value being an increased awareness of any hazards present.

		First aid injury/ illness	Minor injury/ illness	'3 day' injury/ illness	Major injury/ illness	Fatality/ disabling injury
		1	2	3	4	5
Very likely	5	5	10	15	20	25
Fairly likely	4	4	8	12	16	20
Likely	3	3	6	9	12	15
Unlikely	2	2	4	6	8	10
Very unlikely	1	1	2	3	4	5

△ Table 2 Risk assessment.

Apply this skill by doing Activities AIS1, AIS2.

S11 SI UNITS AND eV

SI stands for Système International. It is a system of units constructed from just seven base units and is used by the scientific profession worldwide. The seven base units are in Table 3, and the method for defining the value of each unit is clearly established, with the exception of the kilogram. Currently all prototype kilograms are copies of the International Prototype Kilogram kept at Sèvres near Paris but it is now known that there is measurable divergence between them. The General Conference on Weights and Measures has agreed that the kilogram should be redefined in terms of the Planck constant.

Unit	Unit symbol	Quantity name
metre	m	length
kilogram	kg	mass
second	s	time
ampere	A	electric current
kelvin	K	thermodynamic temperature
mole	mol	amount of substance
candela	cd	luminous intensity

△ Table 3 Base units in the SI system

All the units you will encounter are derived from a combination of the seven base units. For example, electrical conductance is measured in siemens (symbol S). In terms of base SI units this quantity is $kg^{-1}\,m^{-2}\,s^3\,A^2$. Note that where a unit is named after a scientist the unit symbol is capitalised, but when the unit name is used in text it is lower case. For example "the newton, N, is named after Newton".

The unit of energy, the joule (J), is N m in base units. For many purposes the joule is too large a unit. In atomic and nuclear physics the **electron volt**, eV, is used. Recall that the volt is joules per coulomb, p.d.(V) = energy (J)/charge (C). Therefore, potential difference multiplied by charge is the amount of energy associated with the charge. When a single electron moves across a potential difference of one volt it gains or loses one electron volt (1 eV) of energy, where e is the charge on one electron,

$$1\ eV = \text{charge} \times \text{potential difference} = 1.6 \times 10^{-19}\,C \times 1\,V = 1.6 \times 10^{-19}\,J$$

To turn eV into joules multiply by 1.6×10^{-19}

To turn joules into eV divide by 1.6×10^{-19}

Apply this skill by doing Activities A2, A4, A7, A9, A10, A11.

S12 WRITING FOR YOUR INTENDED AUDIENCE

Writing for your intended audience is all about style. Good communication is about knowing your audience and exactly what the purpose of the writing is. It may be descriptive or persuasive. You may simply be providing a set of instructions. Your audience may be the general public, young people or stakeholders in an industry. Find the appropriate style, whether it be bullet points or making a political statement using ethical arguments.

In an examination you can probably assume that your intended audience is a group of scientists or an intelligent fellow-student of your own age. You should therefore avoid the use of popular language and probably use the third person. For example, *I made a big mistake with the force of the Helmholtz rings* might be acceptable in some situations, but in an examination you should probably have written *An error was made when calculating the field due to the Helmholtz coils.* This second sentence uses the third person, *field* instead of *force* and *coils* instead of *rings*. It is clearly addressed to a scientific audience. This book is addressed to you and so you will find the second person has often been used.

Generally, avoid the use of very informal language – the sort of language you might use when talking to friends. The use of expressions such as "bursting with energy" instead of "moving at high speed" would not be appropriate.

Apply this skill by doing Activities A1–A15.

S13 ENSURING MEANING IS CLEAR

At various points in your course, you will be assessed for the quality of your written communication. To ensure the meaning of your writing is clear, it's important to ensure that your spelling, punctuation and grammar are accurate. You may create a negative impression by writing *there* when you mean *their* or *its* when you mean *it's*.

You cannot make your meaning clear unless your sentence construction is accurate. Your best weapon to achieve clarity is the short sentence. When your sentence is short, say 10 to 20 words, you can avoid having to decide about commas and other punctuation. Use the full stop. For example, a student writes:

> *Do not connect neither wire to the earth or supply wire.*

This may possibly be comprehensible to an examiner with the time to work out what you mean, but you will lose marks because the sentence makes little sense and a hurried examiner may not give you the benefit of the doubt. The student meant to say:

> *Do not connect either wire to the earth or supply wire.*

Many errors in communication are caused by making unit mistakes. *The student heated the water to 80 K* is obviously wrong. Writing m s^{-1} when m s^{-2} is meant may lose you marks.

Apply this skill by doing Activities A1–A15.

S14 ORGANISING INFORMATION CLEARLY AND COHERENTLY

You should develop your skills in tabulating data. Make sure that you give units and headings and keep the appropriate number of significant figures.

Organise equation writing so that equations are written below each other with the = signs lining up if possible. Do not write muddled statements such as:

$$s = ut + \tfrac{1}{2}at^2 = \tfrac{1}{2}at^2 = 20 = 20 \times 5 = 100 \text{ W}$$

when you mean:

$$s = ut + \tfrac{1}{2}at^2 = \tfrac{1}{2}at^2 \text{ when } u = 0$$

$$a = 10 \text{ so}$$

$$s = \tfrac{1}{2} \times 10 \times 4 = 20 \quad \text{when } t = 2$$

work done $= 20 \times 5 = 100 \text{ W}$

You have arrived at the same answer but the first statement is full of illogical *equalities*. If you make a mistake it's hard to sort out what you have done and even harder for the examiner to award marks for method.

When writing about a technique or method it may be appropriate to use bullet points. Even if it's not appropriate, you may find it helpful to use bullet points first, then write your passage and cross out the bullet points tidily. This would be perfectly acceptable.

Feel free to use diagrams or informal sketches when explaining or describing. It is very useful to be able to refer to a labelled part of your sketch and helps the examiner to follow your reasoning. When you are asked to sketch a graph be sure to mark intercepts and label axes.

Apply this skill by doing Activities A1–A15.

S15 USING SPECIALIST VOCABULARY

You can be as technical as you wish but always pay critical attention to detail. Look at the way technical language is used in a newspaper and compare it with scientific writing. For example, force, energy and power are often confused in newspapers but not in scientific writing. *Domestic irons use a lot of energy* suggests that irons are wasteful when in fact they use comparatively little energy – they are powerful but are only on for a short time. Similarly, newspapers will refer to the weight of an object but never to its mass. *The weight of the book is 5 kg* is simply wrong in physics writing.

Your use of technical vocabulary must be accurate, but so must your description of processes and relationships. If you write *The induced voltage is proportional to the flux* instead of *the rate of change of flux*, the examiner will think that you have not understood the concept of induction.

Apply this skill by doing Activities A1–A15.

S16 TACKLING EXAM QUESTIONS

Always pay special attention to the command word near the beginning of a question. In particular, the words *state, give, name* all mean write down, so there is no need to do any calculating and this can save you a great deal of time. *Calculate* requires that you show your working as marks are often awarded for method, and you must give the units in the final answer. If the units are provided you can use this information to help you find or check a method.

Discuss requires that you write a few sentences to explain an application of physics or explain a given situation using principles of physics. Remember to keep sentences as short as possible. These questions are often marked for quality of written communication (QWC) but you can also use diagrams or sketches if you find that helpful.

Look carefully at the key words used in your awarding body specification to suggest the depth at which you are expected to study.

You may be asked to recall facts and terminology:

Define	Give the precise meaning of a word, phrase, or physical quantity.
Draw	Use a pencil, ensure clarity (use a ruler if a line is supposed to be straight), add labels and, perhaps, annotations.
Label	Add labels to a given diagram.
List	Give a sequence of names or other brief answers with no explanation.
State	Give a specific name, value or other brief answer without explanation or calculation.

You may be asked to apply concepts and show that you can use scientific terminology:

Annotate	Add brief notes to a diagram or graph.
Apply	Use an idea, equation, principle, theory or law in a new situation.
Calculate	Find a numerical answer showing the relevant stages in the working (unless instructed not to do so).
Describe	Give a detailed account of a technique or method.
Distinguish	Give the differences between two or more different items.
Estimate	Find an approximate value for an unknown quantity.
Identify	Find an answer from a given number of possibilities.
Outline	Give a brief account or summary.

You may be asked to analyse or evaluate methods, explanations or hypotheses:

Analyse	Interpret data to reach conclusions. The data may be in graphical form, in which case analyse means finding gradients, intercepts, maxima and/or minima.
Comment	Give a judgment based on a given statement or result of a calculation.
Compare	Give an account of similarities and differences between two (or more) items, referring to both/all of them throughout.
Construct	Represent or develop in graphical or diagrammatic form.
Deduce	Reach a conclusion from the information given. Always study the information given for clues!
Derive	Manipulate a mathematical relationship to give a new equation or relationship.
Design	Produce a plan, simulation or model.
Determine	Find the only possible answer.
Discuss	Give an account including, where possible, a range of arguments for and against the relative importance of various factors, or comparisons of alternative hypotheses.
Evaluate	Assess the implications and limitations of a method or explanation.
Explain	Give a detailed account of causes, reasons or mechanisms.
Predict	Give an expected result.
Show	Give the steps in a calculation or derivation.
Sketch	Represent by means of a graph showing a line and labelled but unscaled axes and with important features (for example, intercept) clearly indicated.
Solve	Find the value of a variable in an equation using algebraic and/or numerical methods. Solve for V means make V the subject and find a value for it.
Suggest	Propose a hypothesis or other possible answer.

Apply this skill by doing Activities A1–A15.

S17 FRACTIONS

A fraction is simply a ratio of two numbers or algebraic expressions. It is one number or algebraic expression divided by another. The bottom is the **denominator** and the top is the **numerator**. The fraction remains unchanged if the numerator and denominator are multiplied by the same number or algebraic expression. For example,

$$\frac{3}{5} = \frac{3 \times 2}{5 \times 2} = \frac{6ab}{10ab}$$

Dividing two fractions

When dividing one fraction by another remember that division is simply multiplication by the multiplicative **inverse**. For example $\frac{3}{7}$ implies $3 \times (\frac{1}{7})$. The inverse of a fraction is simply the fraction upside down. Below, $\frac{3}{5}$ is divided by $\frac{2}{5}$, so $\frac{2}{5}$ becomes $\frac{5}{2}$ and then multiplies $\frac{3}{5}$.

$$\frac{\frac{3}{5}}{\frac{2}{5}} = \left(\frac{3}{5}\right) \times \left(\frac{5}{2}\right) = \frac{15}{10} = 1.5$$

The rule when dividing two fractions is *turn the bottom fraction upside down and multiply*.

Remember that $\frac{0}{1} = 0$ but $\frac{1}{0} = \infty$. **Infinity** (denoted by ∞) is not a real number so dividing by 0 is forbidden.

Percentage

Five **per cent** of 25 or 5% of 25 means $\left(\frac{5}{100}\right) \times 25 = 1.25$ because 5% means $\left(\frac{5}{100}\right)$.

A 100% increase therefore means a doubling.

To turn any fraction into a percentage simply multiply the fraction by 100%:

$$\left(\frac{3}{5}\right) \times 100\% = 60\%$$

And of course, multiplying by 100% is multiplying by one!

Apply this skill by doing Activities A6, A13, A15.

S18 EXPONENTS

10^3 is a **power** of 10. Sometimes this is read as 10 to the power of 3, or 10 cubed: 3 is the **exponent** or index and 10 is the **base**. It means simply $10 \times 10 \times 10$.

Use the pattern in the box to see that negative exponents mean *take the reciprocal*.

To multiply two powers with the same base simply add the exponents. When dividing two powers with the same base, subtract the exponents.

For example $10^4 \times 10^{-7} = 10^{-3}$

and $10^4 / 10^{-7} = 10^{11}$

$10^3 = 10 \times 10 \times 10$	
$10^2 = 10 \times 10$	
$10^1 = 10$	
$10^0 = 1$	
$10^{-1} = 1/10$	$= 0.1$
$10^{-2} = 1/100$	$= 0.01$
$10^{-3} = 1/1000$	$= 0.001$

$$a^{-p} = 1/a^p$$
$$a^p \times a^m = a^{p+m}$$
$$a^p / a^m = a^{p-m}$$

Fractional exponents

Using the exponent multiplying rule, it follows that $10^{\frac{1}{2}} \times 10^{\frac{1}{2}} = 10^1 = 10$

$$a^{\frac{p}{q}} = \sqrt[q]{a^p}$$
$$(a^p)^q = a^{pq}$$

This implies that $10^{\frac{1}{2}}$ is the square root of 10. This is true for other fractional powers so that $10^{\frac{2}{3}}$ means the cube root of 10 squared.

Apply this skill by doing Activities A1–A15.

S19 CHANGING THE SUBJECT OF AN EQUATION

You will frequently be asked questions in which you will have to change the subject of an equation. When learning how to do this it helps to know the names of parts of an equation. Taking one you may have seen before (from the equations of **uniform motion**):

$$s = ut + \frac{1}{2}at^2$$

The *variables* are s, u, t and a. At the moment s is the *subject* of the equation and we would like to make u the subject. The *terms* are s, ut and $\frac{1}{2}at^2$, while $\frac{1}{2}$ is a **coefficient** and 2 is an exponent. The right-hand side, $ut + \frac{1}{2}at^2$, is an *expression*.

The general rule is that you can do anything you like to an equation so long as you do the same thing to both sides, but you may not divide both sides by zero. You want to get u on its own on the left-hand side.

Since the = sign means that both sides are numerically equal you can write the left-hand side or the right-hand side first so that:

$$ut + \frac{1}{2}at^2 = s$$

Now subtract the term $\frac{1}{2}at^2$ from both sides:

$$\left(ut + \frac{1}{2}at^2\right) - \frac{1}{2}at^2 = s - \frac{1}{2}at^2$$

So that:

$$ut = s - \frac{1}{2}at^2$$

You can now divide both sides by t and you will note that the left-hand side now leaves u as the subject of the equation. When dividing both sides by a term or expression, take care to divide into each term on both sides so that:

$$\frac{ut}{t} = \frac{s}{t} + \left(\frac{1}{2}at^2\right)\bigg/ t \qquad \text{... dividing each term on both sides by } t$$

$$u = \frac{s}{t} + \frac{1}{2}at^2 \qquad \text{... } u \text{ is now the subject of the equation.}$$

As another example, make g the subject of the equation in:

$$T = 2\pi\sqrt{\frac{l}{g}}$$

Square both sides:

$$T^2 = 4\pi^2\frac{l}{g}$$

Multiply both sides by g so that g cancels on the right-hand side and appears on the left:

$$gT^2 = 4\pi^2 l$$

Some people call this cross-multiplying. You can only do this when there is just one term on each side. Now divide both sides by T^2 (or cross-multiply by $1/T^2$) so that g is alone on the left-hand side:

$$g = \frac{(4\pi^2 l)}{T^2}$$

Apply this skill by doing Activities A4, A5, A6.

S20 UNIT ANALYSIS

Measured quantities have units. Units can be handled algebraically. This means that you can determine the units of quantities calculated from measurements. For example, density = mass/volume. Volume has units of m^3 being a length times a length times a length, or length cubed. Therefore, density has units kg/m^3 or $kg\ m^{-3}$. Note that this should be read as kilograms per cubic metre, not kilograms per metre cubed.

Some calculated quantities have special names, for example as $F = ma$, F has units of $kg\ m/s^2$ or $kg\ m\ s^{-2}$. Note the use of s^{-2} to mean $1/s^2$ or per second squared. The compound unit $kg\ m\ s^{-2}$ is given the special name of the newton or N.

Each side of a physics equation is the same numerically and has the same units. This means that each term in an equation must have the same units. For example, in $s = ut + \frac{1}{2}at^2$ each term must have the unit metres, as displacement on the left is in metres. This means $\frac{1}{2}at^2$ must also be in metres. The coefficient $\frac{1}{2}$ has no unit. a is in $m\ s^{-2}$ and t^2 is in s^2 so:

$$m\ s^{-2} \times s^2 = m$$

Note the risk of confusion here with the symbol for displacement, s, and the symbol for seconds, which is also s. In print, variables are italic, but units are not. Note also that compound units have spaces, so that m s is metre × second, whereas ms is millisecond.

The manipulation of units in this way is very powerful. It enables you to find the unit for constants such as **resistivity** ρ in the equation or formula $R = \rho(l/A)$. The term $\rho(l/A)$ must have the unit Ω, because that is the unit of R.

Making ρ the subject of the equation, you have:

$$\rho = R \times \frac{A}{l}$$

$$\text{unit of } \rho = \Omega \times \frac{m^2}{m} = \Omega\ m$$

Some quantities have no unit. For example, efficiency = work out / energy in. Since work and energy both have the unit joule, efficiency has no unit and is simply a number (in this case always less than 1, unless expressed as a percentage).

Apply this skill by doing Activities A9, AIS1, AIS2.

S21 SOLVING EQUATIONS

For the equation

$$T = 2\pi\sqrt{\frac{l}{g}}$$

solving for **g** means making g the subject of the equation and finding its value with the associated unit. You will meet this equation when investigating **pendulums**; T is the period of the pendulum in seconds, l is the length of the pendulum in metres and π is just a number. Rewriting the equation with g as the subject, you have:

$$g = \frac{(4\pi^2 l)}{T^2}$$

Values for π, l and T must be substituted into the right-hand side and the term then evaluated, carrying out the calculations in the correct order. In the example given, the order does not matter as each variable is multiplied or divided by its neighbours. However, when using your calculator you must be careful to square only π and T, so be sure that you know how your calculator carries out operations.

g in the equation above has units given by (unit of l divided by unit of T^2) or (m s^{-2}), which you will recognise as the unit of acceleration. Note that acceleration can also be written as $a = F/m$ by making a the subject of the equation $F = ma$.

This means that acceleration also has units N/kg or N kg^{-1}. The unit N kg^{-1} is identical to the unit m s^{-2}, which means that g, the acceleration due to gravity, can also be described as the force per unit mass, or gravitational field strength, k, in N kg^{-1}, newtons per kilogram.

To avoid making mistakes it is good practice to make sure that when you substitute values, all units are SI units. In this way any solution will automatically be in an SI unit.

Apply this skill by doing Activities A3, A6, A8, A12, A14.

S22 SIMULTANEOUS EQUATIONS

When solving for a variable in an equation, the values of all the other variables must be known. An equation that relates two variables does not have a unique solution if both variables are unknown. However, another equation relating the same two variables may coincide with the first equation at one point. Graphically, two lines will intersect somewhere if the gradients are different. That point of intersection is the "simultaneous solution". The two equations are known as a pair of **simultaneous** equations. For example, if two teas and one coffee cost £1.70 and three teas and two coffees cost £2.80, what is the cost of one tea and one coffee? The two equations are here solved by linear combination or elimination. Equation **a** is multiplied by 2 and equation **b** is subtracted from it.

$2t + c = 1.7$	**a**
$3t + 2c = 2.8$	**b**
$4t + 2c = 3.4$	**a** × 2
$t = 0.6$	(**a** × 2) − **b**
$c = 1.7 - 2t = 0.5$	from **a**

The two equations have been graphed (see Fig 9) and the intersection found, (0.5, 0.6). So one tea and one coffee costs $0.6 + 0.5 = £1.10$.

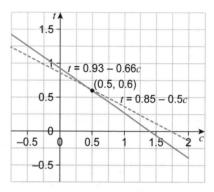

If a unique solution to the pair of simultaneous equations exists then it may also be found by algebraic substitution. Simply choose one variable and make it the subject of the equation. Then substitute the right-hand side into the same variable in the other equation.

For example, a train sets off from a station with an acceleration of 0.5 m s^{-2}. A second train sets off 15 s later from the same station in the same direction with an acceleration of 2 m s^{-2}.

△ Fig 9 Intersection of two equations.

Providing the acceleration is maintained, after how long will the second train catch the first?

$s = \frac{1}{2}at^2$	
$s_1 = \frac{1}{2} \times 0.5 \times t^2$	first train
$s_2 = \frac{1}{2} \times 2 \times (t - 15)^2$	second train

When the second train catches up with the first train, $s_1 = s_2$ and so you can simply equate the right-hand sides. This gives a quadratic equation in t (see S23). Solving for t gives two solutions, 30 or 10 seconds. Since the second train sets off 15 seconds after the first train, the solution must be 30 s, for which $s_1 = s_2 = 225$ m.

Apply this skill by doing Activity A6.

S23 QUADRATIC EQUATIONS

Linear equations, such as $v = at$, are straight lines when graphed. None of the variables is raised to a power. The equation $s = \frac{1}{2}at^2$ has a variable raised to the power 2. This is called a quadratic equation.

In the equation $y = ax^2 + bx + c$ the term with coefficient a shows x as a power 2. This is the general form of a quadratic equation; similarly, $y = ax^3 + bx^2 + cx + d$ is a cubic equation; and so on.

The graphical plot of the quadratic equation $y = ax^2 + bx + c = 0$ is a curve known as a parabola (Fig 10). It is a U-shaped curve which is upright if the coefficient a is positive; that is, it has a minimum value at the bottom of the U. The curve is upside down if a is negative; that is, it has a maximum value at the top of the U. If the U cuts the x-axis (where $y = 0$) then there are two solutions to the equation $ax^2 + bx + c = 0$. If it just touches the x-axis (that is, the x-axis is a tangent to the parabola) then there is just one solution. If it does not cut or touch the x-axis at all then there are no real solutions (they exist in the set of imaginary numbers, which you need not be concerned with in your physics).

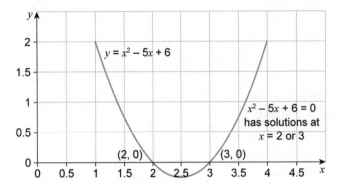

△ Fig 10 Graph of a quadratic equation.

Sometimes a quadratic can be solved by factorisation. For example, $x^2 - 5x + 6 = 0$ is a quadratic and can be rewritten $(x - 2)(x - 3) = 0$ where $(x - 2)$ and $(x - 3)$ are factors of the quadratic expression. This means that either $(x - 2)$ or $(x - 3)$ can equal zero for the equality to be satisfied. There are then two solutions, $x = 2$ or $x = 3$. The equation could also be solved using the quadratic formula:

$$x = \frac{-b \pm \sqrt{b^2 - 4ac}}{2a}$$

which also yields $x = (5 \pm 1)/2 = 3$ or 2.

The equation $s = ut + \frac{1}{2}at^2$ is a quadratic in t that expresses the displacement of an object moving with constant acceleration with time t and initial velocity u. For any value of s there may be two solutions or two values of t that should satisfy the equation. Assume for an object thrown upwards at 20 m s^{-1} that $a = 10$ m s^{-2} downwards.

1. When $s = 20$ m $20 = 20t - 5t^2$ or $5t^2 - 20t + 20 = 0$

   Dividing both sides by 5 and then factorising the expression gives $(t - 2)^2 = 0$ from which we get the single solution $t = 2$. The two solutions are said to be coincident, that is they are the same. At 20 m the projectile has reached the top of its flight.

2. When $s = 15$ m $15 = 20t - 5t^2$ or $5t^2 - 20t + 15 = 0$

   Dividing both sides by 5 and then factorising gives $(t - 1)(t - 3) = 0$ from which $t = 1$ or 3. That means the projectile reaches 15 m on the way up after 1 second and then 2 seconds after that it has passed through the top at 20 m and come back down to the 15 m mark again.

3. When $s = 0$ m $0 = 20t - 5t^2$ or $5t^2 - 20t = 0$

   Dividing both sides by 5 and factorising gives $t(t - 4) = 0$ from which $t = 0$ or 4. The projectile is at ground level when $t = 0$ and at 4 s.

You should check that all three situations give the same answers using the quadratic formula.

What happens when we put $s = 25$ m? The equation will not factorise and the quadratic formula does not provide any real solutions because the part $\sqrt{(b^2 - 4ac)}$ is negative and so the equation only has imaginary solutions. The projectile can never reach the height 25 m with an initial upwards velocity of 20 m s^{-1}.

Note that in the equation $s = ut + \frac{1}{2}at^2$ all the terms contain a **vector** so that the equation is a vector equation. It is important to give a sign to each of s, u and a to signify whether the vector is upwards or downwards. It does not matter which direction you take as positive, so long as you are consistent.

Apply this skill by doing Activity A3.

S24 INVERSE FUNCTIONS

Suppose the area of a rectangular room in a new house is to be 10 m^2, then how does the length vary with the width of the room? You have length × width = 10. Making length the subject, length = 10/width. If the width is small, 10/width is large; and if the width is large, 10/width is small. This means length and width are inversely related. We say length *varies inversely* as width. If you plot length against width (Fig 11), you get a curve **asymptotic** to both the length axis and the width axis.

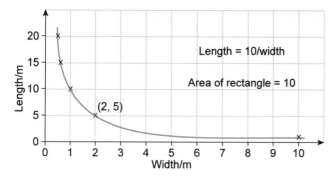

△ Fig 11 Length plotted against width.

If you plot length against 1/width (Fig 12), instead of length against width, the result is a straight line showing proportionality: length ∝ 1/width.

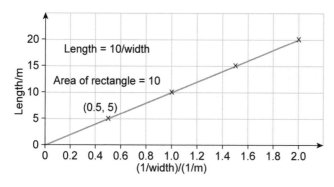

△ Fig 12 Length plotted against 1/width.

The volume of a square **pyramid** is (1/3 × base area × height). The base area is the side of the base squared, and if the volume is fixed to some value, say 20 m³, the height will vary inversely as the square of the side of the base. This produces another graph curve, again asymptotic to both axes (Fig 13). It is difficult to be sure what the relationship is from the sketched curve – is it an inverse or an inverse square relationship?

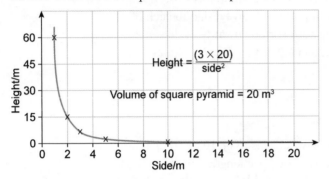

△ Fig 13 Graph of inverse or inverse square relationship.

Again, physicists find straight line graphs helpful, and if height is plotted against the inverse of the side squared you see a perfectly straight line (Fig 14). You can now be sure that this is an inverse square relationship; the height varies inversely as the square of the side of the base. *Straightness* is easily identifiable and therefore the relationship can be written down. It is much easier to make predictions from a straight line than from a curve.

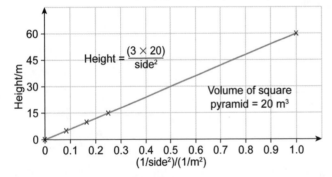

△ Fig 14 Height plotted against 1/side².

Apply this skill by doing Activity A15.

S25 SIGNIFICANT FIGURES

The number of **significant figures** in a measured or calculated quantity is a measure of the significance or precision of the measurement. The number 562 has three significant figures. The number 560 may have three or two significant figures, as it is not possible to know if the trailing zero is significant or not. The number 560.3 has four significant figures and the number 560.30 has five significant figures, as this time the trailing zero has been purposely included. It is a genuine product of the calculation or measurement.

Trailing zeros after the decimal point are always significant.
Trailing zeros before the decimal point may or may not be significant.

Most measurements in advanced physics will be to two or three significant figures. Therefore, calculated results will be to two or three figures. As a simple rule, only retain as many figures in the result as in the least significant measurement. This simple rule applies to examination questions as well. When calculating a result from given data, only quote as many figures as in the least significant quantity.

Apply this skill by doing Activities A2, A11, AIS1, AIS2.

S26 SCIENTIFIC NOTATION

Scientists write down measurements and values in a form that lets the reader always know how many figures are significant. This form is also useful for handling very large or very small numbers. The decimal point is moved so that only one digit is before the point.

The number 560 then becomes 5.6×10^2 or 5.60×10^2 depending on whether the zero is significant or not. 5.60×10^2 has three significant figures as here the trailing zero has been written purposely after the decimal point.

It is meaningless to talk about the number of decimal places of a measurement. Say your height is 1.72 m to three significant figures. This is written to two decimal places. By changing the unit the decimal places are changed. Your height in km is 0.00172, which is still three significant figures but now has five decimal places! More correctly you would write 1.72×10^{-3} km. Written this way the significance of the measurement is clearly the same whether in kilometres or metres. The power of 10 is simply a multiplier (see S38 Estimating).

Apply this skill by doing Activities A1–A15.

S27 GRADIENT OF A STRAIGHT LINE

All straight lines can be written in the form $y = mx + c$ where m is the **gradient** of the line and c is the intercept on the y-axis (see Fig 15). Putting y to zero, the x-intercept is $(-c/m)$. If $m = 0$ the line is horizontal and $y = c$.

The gradient m tells you how y changes as x changes. The slope of a line can be understood in terms of the angle it makes with the x-axis. The tangent of the angle is the gradient. If the gradient m is negative then y decreases as x increases.

If you have plotted data and drawn a straight line you can determine the gradient by selecting the largest sensible x-interval and finding what the corresponding y-interval is. The gradient is then equal to the (y-interval/x-interval). Clearly the larger the interval the greater the number of significant figures and the greater the number of significant figures in the gradient, m.

Note that the unit of the gradient is found from inspecting the units of the y-interval and the x-interval. If the x-interval is in seconds then the gradient is the rate of change of y with time.

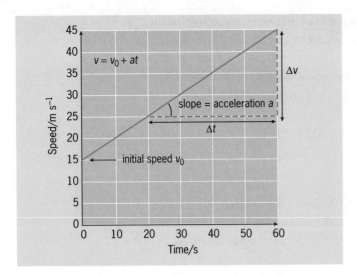

△ Fig 15 Slope and intercept graph of accelerated motion.

Apply this skill by doing Activities A1, A2, A3, A5.

S28 GRADIENT OF A CURVE

The gradient of a curve is continually changing with x. To determine the gradient for some x-value you must draw the tangent to the curve at x (see Fig 16). Then, as for the straight line, determine the gradient of the tangent. Note that a tangent to a curve touches the curve at just one point.

If the gradient of the tangent is zero then you are at a maximum or a minimum point of the curve; this therefore represents a maximum or minimum value of y.

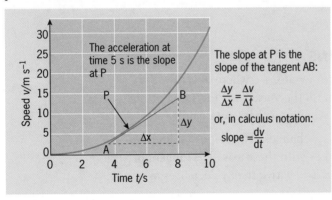

△ Fig 16 Drawing a tangent to a graph.

Apply this skill by doing Activities A3, A6, A15.

S29 AREA UNDER A CURVE

The area under a curve represents a physical quantity found by multiplying the y and x quantities. For example, in a graph of velocity against time you already know that the gradient is velocity/time, which is acceleration in m s^{-2}. The area under the curve has units m s$^{-1} \times$ s, which is simply m. The area has the unit of length, and is the change in displacement of the object over the time interval chosen.

In terms of calculus, the gradient of a function is found by **differentiating**, while the area under the graph of a function is found by **integrating**. For example, differentiating a velocity–time graph gives the acceleration, and integrating an acceleration–time graph gives the average velocity.

The simple way to estimate the area under a graph is to count squares, and even in many advanced physics courses this will be sufficient (see Fig 17).

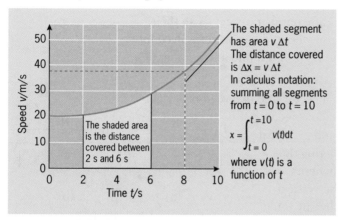

The shaded segment has area $v \, \Delta t$
The distance covered is $\Delta x = v \, \Delta t$
In calculus notation: summing all segments from $t = 0$ to $t = 10$

$$x = \int_{t=0}^{t=10} v(t) \, dt$$

where $v(t)$ is a function of t

△ Fig 17 Calculating an area under a curve.

Apply this skill by doing Activities A3, A9, A13.

S30 VECTORS

Vectors are mathematical objects that require more than one number for their complete description. In physics there are many quantities that have both magnitude and direction, and can be treated as vector quantities using vector algebra. Typically, force, acceleration, velocity and momentum are vectors, and can be represented as arrows that have a length representing their size and a direction. The two diagrams shown in Fig 18 are equivalent. Fig 18(a) shows vector $\underline{a}$ followed by vector $\underline{b}$ being equal to vector $\underline{c}$. In Fig 18(b) vector $\underline{b}$ has been displaced to show it originating from the same point as vector $\underline{a}$. Two vectors are equal if they are parallel and have the same direction, so that in both diagrams the vector $\underline{b}$ is the same.

If $\underline{a}$ and $\underline{b}$ are displacements it is clear from the first diagram that displacement $\underline{a}$ followed by displacement $\underline{b}$ is displacement $\underline{c}$. If $\underline{a}$ and $\underline{b}$ are forces then the second diagram is easier to understand, as both $\underline{a}$ and $\underline{b}$ would be pulling from the same point in their respective directions and would then be equal to force $\underline{c}$.

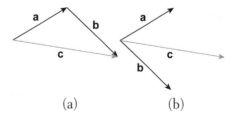

Vector $\underline{c}$ is said to be the **resultant** of vectors $\underline{a}$ and $\underline{b}$. Vectors are usually represented in bold type in print or with a dash underneath or an arrow on top.

△ Fig 18 Equivalent vector diagrams.

In terms of coordinates, (x_1, y_1) could describe a vector from the origin to that point. Another vector (x_2, y_2) is added to the first. The resultant is simply $\{(x_1 + x_2), (y_1 + y_2)\}$.

Vectors can also be subtracted. Since $\underline{a} + \underline{b} = \underline{c}$, it follows that $\underline{a} = \underline{c} + (-\underline{b}) = \underline{c} - \underline{b}$. Care should be taken here to understand that $\underline{c} - \underline{b}$ means $\underline{c} + (-\underline{b})$.

Apply this skill by doing Activities A7, A14.

S31 RESOLVING VECTORS

In Fig 19(a) $\underline{d} = \underline{a} + \underline{b} + \underline{c}$. You could draw any number of vectors so their resultant was equal to $\underline{d}$. You can say that the vector $\underline{d}$ has been resolved into the vectors $\underline{a}$, $\underline{b}$ and $\underline{c}$. Clearly vector $\underline{d}$ can be resolved into any number of vectors in any number of directions.

It is often very useful, however, to resolve a vector into two vectors at right angles to each other. Fig 19(b) shows the same vector $\underline{d}$ resolved into a horizontal component, $\underline{H}$ and a vertical component, $\underline{V}$. This is useful when considering the velocity vector of a projectile as two components, one being affected by the force of gravity in the vertical direction and the other being unaffected by any force (except air resistance) in the horizontal direction.

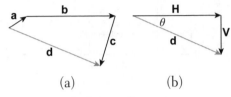

(a)   (b)

△ Fig 19 Resolved vector diagrams.

Using trigonometry for the magnitudes of the vectors:

$$H = d \cos \theta$$

$$V = d \sin \theta$$

and using Pythagoras:

$$d = \sqrt{(H^2 + V^2)}$$

The angle between vector $\underline{d}$ and the horizontal is given by:

$$\tan \theta = V/H$$

You are unlikely to encounter exam questions that require anything more complicated than resolving a vector into two perpendicular directions; this technique is so useful it is worth committing the above equations and Fig 19 to memory.

Apply this skill by doing Activity A7.

S32 TRIGONOMETRY

The word trigonometry has Greek roots and means *triangle measuring*. The right-angled triangle shown in Fig 20 has been labelled traditionally. The *hypotenuse* is c; b is *opposite* to the angle θ and a is *adjacent* to θ. The dotted lines show a similar triangle made by extending the sides c and a. All the angles in the new triangle remain unchanged. The ratios of all the sides also remain unchanged. These ratios are the trigonometric ratios.

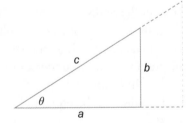

△ Fig 20 Right-angled triangle.

There are six ways of calculating these ratios and these are given in the box. The first three are usually shortened to sin, cos and tan and are frequently used. Once the sides are known, the angles are also known.

Check that your calculator is in "degree" mode and look up sin 30. You should get 0.5. Suppose the ratio b/c is 0.8660 (four significant figures). Enter this ratio in your calculator and find $\sin^{-1}$. This means "the angle whose sin is" and is written as inverse sin or arcsin. It simply means "going backwards" and must not be confused with $1/\sin\theta$. Pressing $\sin^{-1}$ in this case should give you 60.00 to 4 significant figures.

SOHCAHTOA is a popular way of remembering which sides to use for the ratios:

Sin **O**pposite over **H**ypoteneuse **C**os **A**djacent over **H**ypoteneuse **T**an **O**pposite over **A**djacent

sine θ	$= b/c$
cosine θ	$= a/c$
tangent θ	$= b/a$
cosecant θ	$= c/b$
secant θ	$= c/a$
cotangent θ	$= a/b$

These ratios are also known as the circular functions because they can be generated by considering the rotating radius vector of a circle (see Fig 21). If the length of the radius vector is always positive you can determine the signs of the ratios when the vector is in the different quadrants of the circle. All the ratios are positive in the first quadrant but in the second only sin is positive, in the third only tan is positive and in the fourth quadrant only cos.

The sin of 150° is found by subtracting it from 180° so that $\sin 150° = \sin 30° = 0.5$.

Cos 120° = $-\cos 60° = -0.5$ as it is in the second quadrant.

It may be useful to learn the trigonometric ratios of some easy angles, shown in Table 4.

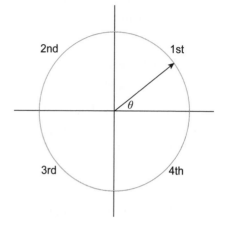

△ Fig 21 Rotating radius vector of a circle.

θ/degrees	0	30	45	60	90
sin θ	0	0.5 (1/2)	0.707 (1/√2)	0.866 (√3/2)	1
cos θ	1	0.866 (√3/2)	0.707 (1/√2)	0.5 (1/2)	0
tan θ	0	0.577 (1/√3)	1	1.732 (√3)	∞

△ Table 4 Trigonometric ratios.

There are two trigonometric equations, known as identities, which are useful in some areas of physics. These are shown in the box. They are both easily proved using Pythagoras' theorem.

$$\sin^2\theta + \cos^2\theta = 1$$
$$\sin 2\theta = 2\sin\theta\cos\theta$$

Using sin² and cos²

When dealing with ac voltages (potential differences) the voltage is changing all the time, so how is the voltage quoted? Since $V = V_0\sin\omega t$ where V_0 is the maximum value of the voltage, it is easy to see that the average value of V over time is zero since there is much of the sin curve above zero as there is below, but the voltage is only equal to zero at two instants of time in every cycle as shown in the diagram.

By squaring the ac voltage or the ac current all the negative values are made positive. The average is now found by taking the square root, and then the mean of

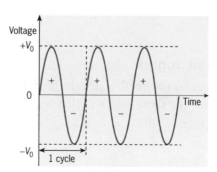

△ Fig 22 Three complete cycles of alternating voltage. The average over a complete number of cycles is zero.

MATHS

the squared voltage. This new curve is a sin² curve and as the maximum value of sin is 1 so the maximum value of sin² is 1. This means that the maximum value of $(V)^2$ is the maximum value of $(V_0 \sin \omega t)^2$ or $(V_0)^2$. The average value of this new curve is simply half the peak value or $(V_0)^2/2$.

The maximum voltage is now $(V_0)^2$ so that taking the square root gets the maximum voltage back to V_0. The mean of the $(V_0)^2 \sin^2 \omega t$ curve is $(V_0)^2/2$. The root of the mean of the squares (rms) is therefore:

$$V_{rms} = \sqrt{[(V_0)^2/2]} = V_0/\sqrt{2}$$

where V_0 is the peak voltage or V_{peak}. Therefore:

$$V_{peak} = \sqrt{2} \times V_{rms}$$

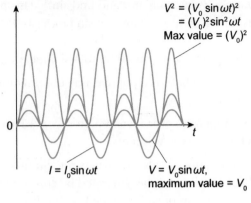

△ Fig 23 The sin² curve is entirely positive.

Mains voltage in the UK is 230 V and this is the root mean square value (rms), so that the peak voltage is $\sqrt{2} \times 230 = 325$ V. Any insulation must be able to withstand 325 V, not 230 V.

The power curve $(I \times V)$ is also a sin² curve so that average power (peak power/2) is $I_{rms} \times V_{rms}$. Check that this is correct.

Apply this skill by doing Activity A6.

S33 RADIANS

The radian (rad) is a natural measurement of angle. It is the only physical unit derived from an idealised geometrical shape, the circle. Consider the sector of the circle shown in Fig 24, where r is the radius of the circle and s is the length along the section of circle perimeter. The angle θ in radians is then given by

$$\theta = s/r$$

The symbol ^{c} is sometimes used to signify radian measure, as opposed to degrees °. If $s = 2\pi r$, $\theta = 2\pi^c$. Hence $360° = 2\pi^c$.

△ Fig 24 Sector of a circle.

When using your calculator, ensure that you can change from radians to degrees as needed.

Radian measure is essential when dealing with the equations of simple harmonic motion. When differentiating or integrating trigonometric functions, radian measures must be used.

Small angles

Note that for small angles the arc length s and a perpendicular to r have almost the same length so that $\sin \theta \approx \tan \theta \approx \theta$ for small values of θ in radians.

Apply this skill by doing Activity A12.

S34 ANGULAR MOTION

When a body rotates, it completes a number of revolutions per second. Each revolution is one cycle, and the number of revolutions per second is a frequency measured in hertz (Hz). Five revolutions per second is 5 Hz.

Knowing the angular frequency, f, it is easy to determine the angular velocity, $\theta/t = \omega$, in radians per second by simply multiplying the angular frequency by 2π, the number of radians in one revolution:

$\omega = 2\pi f$ in radians per second.

Angular velocity must not be confused with tangential velocity. Suppose a wheel, radius 0.25 m, is completing five revolutions per second or has an angular frequency of 5 Hz. Its angular velocity ω is $10\pi^c$ s^{-1}. The tangential velocity v_t at the rim of the wheel is $v_t = r\omega$ in m s^{-1} so $v_t = 7.9$ m s^{-1}. Note that tangential velocity is proportional to the radius for constant ω.

Simple harmonic motion

Solutions to the equation $y = A \sin\theta$ yield the displacement of a **simple harmonic oscillator** for values of θ. Substituting for $\theta = \omega t$, and $\omega = 2\pi f$, gives:

$y = A \sin\theta = A \sin\omega t = A \sin(2\pi f t)$

This is a wave equation with y as a function of t, the time. Note that this only gives y as a function of t and is therefore the displacement of the wave at one point, namely $x = 0$. Note the connection between the constant ω used in SHM and ω meaning angular velocity.

Solutions to $y = A \sin\theta$
y is positive for $(2n + 1)\pi > \theta > 2n\pi$
y is negative for $(2 + 1)\pi < \theta < 2n\pi$
for integer n.
For $\theta = (4n + 1)\pi/2$, $y = A$
For $\theta = (4n + 3)\pi/2$, $y = -A$

$y = A \sin\theta$ is graphed in Fig 25 and two consecutive equal values of y marked. The sin of the two marked angles is the same so that $\sin\theta = \sin(180° - \theta)$.

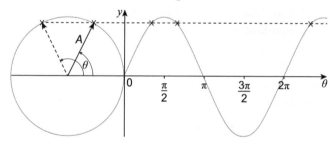

△ Fig 25 Graph for $y = A \sin\theta$.

Apply this skill by doing Activities A8, A12, A14.

S35 EXPONENTIAL CHANGE

Any function of the form $y = a^x$ is an exponential function; we say y is changing exponentially. Recall that x here is the **exponent** of a, and a^x is a power with **base** a.

For example, $y = 2^x$ doubles for every integer increase of x. Note that the graph passes through the point $(0, 1)$. Any base can be chosen and all of them will pass through $(0, 1)$. The graph shown in Fig 26 is an exponential increase; $y = 2^{-x}$ will decrease exponentially, also passing through $(0, 1)$ and approaching the x-axis asymptotically.

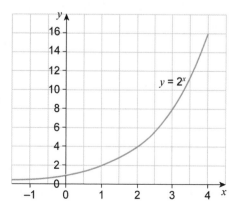

△ Fig 26 Graph of exponential function.

One base is of special interest. If a tangent is drawn to the graph curve at some point, the rate of increase of the function can be measured. The rate of increase of 2^x is always less than the value of 2^x itself; the rate of increase of 3^x is always greater than the value of 3^x. Somewhere in between 2 and 3 there is a base which has the property that the rate of increase of (base)^x is exactly equal to (base)^x itself. That number is 2.71 (to 3 figures), denoted by e. So $y = e^x$ has the unique property of increasing at a rate equal to its own size. For this reason e^x is known as *the* exponential function, and it occurs in many physical situations. In Fig 27 the tangent at $x = 1.95$ is drawn and has a gradient of 7; the value of $y = e^{1.95}$ is also 7.

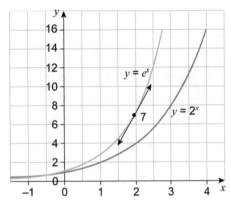

△ Fig 27 The gradient of $e^x = e^x$.

Quantities that change with time at a rate equal to, or proportional to, their size or value, are changing exponentially. This means that the proportional or fractional change for equal intervals of time is a constant. Populations, for example, change exponentially if not limited by predators or lack of resources. Of interest is the time it takes to double; human populations have shown a doubling time of about 40 years. Bacteria can have doubling times from 20 minutes to several hours, depending on environment.

What is needed is an equation that can be solved. For a population P:

$$\frac{\Delta P}{\Delta t} = kP$$

which says that the rate of increase of P is proportional to P. The constant of proportionality is k. This equation in calculus notation is a first-order linear differential equation. To find solutions, the equation can be integrated to give P as a function of t:

$$P = P_0\, e^{kt}$$

where P_0 is the population at time $t = 0$.

Apply this skill by doing Activities A13, AIS2.

S36 LOGARITHMS

A logarithm is another word for **exponent**. For example:

3 is the logarithm of 10^3 to the base 10.

5 is the logarithm of e^5 to the base e.

The base of *natural logarithms* is e. These logarithms are usually denoted ln. Logarithms to base 10 are denoted log or $\log_{10}$.

The three laws of logarithms in the box will be useful when analysing logarithmic equations resulting from exponential functions. The rules apply to ln functions as well as log functions.

$$\log a + \log b = \log (ab)$$
$$\log a - \log b = \log (a/b)$$
$$\log (a^n) = n \log a$$

Note that $\log (10^a \times 10^b) = \log 10^{(a+b)} = a + b = \log 10^a + \log 10^b$. The rules for logs come directly from the rules for exponents (see S18).

In the equation for population growth, $P = P_0\, e^{kt}$, kt is the natural logarithm of e^{kt}. Rewriting the equation:

$$\frac{P}{P_0} = e^{kt}$$

you can now take logarithms of both sides so that:

$$\ln\left(\frac{P}{P_0}\right) = kt$$

If $P = 2P_0$ then t is the doubling time so that $\ln (P/P_0) = \ln 2 = kt$. Since the doubling time is 40 years, $k = \ln 2/40$. Therefore $k = 0.017$ year^{-1} and the population is growing at 0.017 of P per year.

Apply this skill by doing Activity AIS2.

S37 LOGARITHMIC GRAPHS

Quantities that change exponentially, like sound levels, earthquake intensities or number of radioactive nuclei, are often difficult to graph because they grow too large or too small too quickly. In these situations it can be useful to draw a logarithmic scale (Fig 28). Instead of increasing linearly or arithmetically as 2, 4, 8, 10, etc., the scale will increase geometrically as a power series, 2, 4, 8, 16, etc., that is, 2^x or more: usually 10^x or e^x.

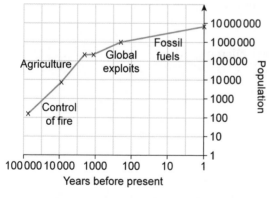

△ Fig 28 Graph using a logarithmic scale.

On this graph both scales are logarithmic, with the increases as powers of 10. It simply would not be possible to represent this data using arithmetic scales.

Taking the function $\ln\left(\frac{P}{P_0}\right) = kt$, recall that dividing powers means subtracting the exponents so that the equation can be written:

$$\ln P - \ln P_0 = kt$$

$$\ln P = kt + \ln P_0$$

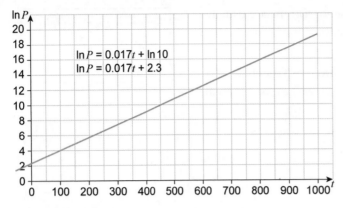

△ Fig 29 Straight line gradient of ln P against t.

A graph of ln P against t is a straight line of gradient k with intercept on the ln P-axis of ln P_0 (see Fig 29). If the starting population is 10, ln 10 = 2.3 and the gradient is 0.017.

The y-axis is ln P so that the scale is arithmetic. Note that the time scale is also arithmetic. The logarithms of exponentially increasing quantities can be plotted on arithmetic (or linear) scales, or the exponentially increasing quantities can be plotted on logarithmic scales.

Suppose you want to know the population after 1000 years, ln P = 19.3 from the graph. You need the number P whose logarithm to the base e is 19.3. On your calculator type 19.3 and inverse ln (or you may have an e^x button; $P = e^{19.3}$). You should get about 2.41×10^8.

Apply this skill by doing Activity AIS2.

S38 ESTIMATING

Estimating in physics is a skill that helps you to know if you are roughly correct in your thinking. In estimating, it is useful to know the values of some quantities as powers of 10, for example in Table 5:

Wavelength of visible light	10^{-6} m or 1 micron (μm)
Gravitational constant G	10^{-10} N m^2 kg^{-2}
Electric field constant k ($k = 1/4\pi\varepsilon_0$)	10^{10} N m^2 C^{-2}
Atomic diameter	10^{-10} m
Nuclear diameter	10^{-15} m
π^2	10

△ Table 5 Some useful values.

These are simply rough indicators. Visible light has a range of wavelengths from 0.4 to 0.7×10^{-6} m, and atomic diameters depend on the particular atom chosen.

• When calculating by substitution in a formula, ensure that all units are correct and that no centimetres or grams have been accidentally left in.
• Write your values in scientific notation, as powers of 10. Calculate the powers of 10 separately from calculating the coefficients of the powers.

For example, you have to calculate the resistivity of aluminium by measuring the resistance of a piece of wire, 41 cm long and 2.4 mm in diameter. The resistance is $2.3 \times 10^{-3} \, \Omega$. Write out the formula and convert measurements to SI.

$$\rho = R\left(\frac{A}{l}\right)$$

$$\rho = \frac{2.3 \times 10^{-3} \, \Omega \times 3.14 \times (1.2 \times 10^{-3} \, \text{m})^2}{4.1 \times 10^{-1} \, \text{m}}$$

$$\rho = \frac{2.3 \times 3.14 \times (1.2)^2}{4.1} \times \frac{10^{-3} \times 10^{-6}}{10^{-1}} \, \Omega \, \text{m}$$

Immediately, the power of 10 is 10^{-8}, from which you can tell that you are roughly correct (assuming you know what the resistivity of aluminium should be). You can then enter the coefficients into the calculator without worrying about the powers of 10, giving 2.54 to three figures.

$$\rho = 2.54 \times 10^{-8} \, \Omega \, \text{m}$$

Apply this skill by doing Activities A11, A13.

Skills to Activities table

This table lists the activities that practise each skill.

Skill type	Skill name	Links to activities
Working Scientifically	1 From hypothesis to theory	A1, A7, A10
	2 Solving physics problems	A3, A10
	3 Energy conservation	A6, A10, A11
	4 Momentum conservation	A7
	5 Free body diagrams	A4, A14
	6 Flux and field lines	A12, A15
	7 Uncertainty (absolute)	AIS1, AIS2
	8 Uncertainty (percentage)	AIS1, AIS2
	9 Systematic and random error	AIS1, AIS2
	10 Risk assessment	AIS1, AIS2
	11 SI units and eV	A2, A4, A7, A9, A10, A11
Quality of Written Communication	12 Writing for your intended audience	A1–A15
	13 Ensuring meaning is clear	A1–A15
	14 Organising information clearly and coherently	A1–A15
	15 Using specialist vocabulary	A1–A15
	16 Tackling exam questions	A1–A15
Maths	17 Fractions	A6, A13, A15
	18 Exponents	A1–A15
	19 Changing the subject of an equation	A4, A5, A6
	20 Unit analysis	A9, AIS1, AIS2
	21 Solving equations	A3, A6, A8, A12, A14
	22 Simultaneous equations	A6
	23 Quadratic equations	A3
	24 Inverse functions	A15
	25 Significant figures	A2, A11, AIS1, AIS2
	26 Scientific notation	A1–A15
	27 Gradient of a straight line	A1, A2, A3, A5
	28 Gradient of a curve	A3, A6, A15
	29 Area under a curve	A3, A9, A13
	30 Vectors	A7, A14
	31 Resolving vectors	A7

Skill type	Skill name	Links to activities
	32 Trigonometry	A6
	33 Radians	A12
	34 Angular motion	A8, A12, A14
	35 Exponential change	A13, AIS2
	36 Logarithms	AIS2
	37 Logarithmic graphs	AIS2
	38 Estimating	A11, A13

Activities

A1 AGE OF THE UNIVERSE

In 1929, Edwin Hubble produced the first evidence that the Universe had an age. Before that most scientists argued for a static Universe but had difficulty reconciling this with new ideas about entropy. If everything is tending to disorder and heat is always flowing from high to low temperature, then the Universe by now should be at one uniform temperature; this is the heat death of the Universe.

Using large telescopes, Hubble showed that many objects regarded as nebulae (faint clouds emitting light) were in fact gigantic clusters of stars, that is, galaxies. These exist well beyond the stars we can see with our eyes. The stars or constellations we can see belong to our own galaxy, the Milky Way. The galaxy Andromeda is visible to the naked eye and appears very roughly as big as the full Moon but is extremely faint and so very difficult to see. Slipher (1912)

△ Fig 30 The Hubble Space Telescope honoured the huge impact that Edwin Hubble made to astronomy.

had already shown, using spectroscopic data, that nearly all the galaxies had "red shifts". The red shift is the increase in wavelength of spectral lines, so called because visible wavelengths move towards the red end of the spectrum. Hubble's later observations supported the conclusion that the galaxies must be receding from us because of this red shift. Indeed, LeMaitre's idea, in 1927, of an expanding Universe provided one explanation of the observed red shifts of the spiral galaxies. LeMaitre predicted Hubble's Law and estimated a value for what is now called the Hubble constant, before Hubble's observations.

The spiral galaxies are now known to be receding from each other at velocities proportional to their distance apart, implying that they once occupied the same place at the same time, the moment of the Big Bang. These velocities are believed to be caused by the expansion of space itself and not by the galaxies moving at ever-higher speeds towards an unknown boundary of the Universe. As such, the phenomenon of the red shift of the galaxies is known as the cosmological red shift. The light from a galaxy a billion light-years away left a billion years ago and in that time space has expanded and "stretched" the radiation within it so that the wavelengths have increased, i.e. experienced a red shift.

The graph in Fig 31 shows the velocity of recession of galaxies, gathered by looking through a telescope in the direction of the constellations named on the graph, plotted against distance. (See also Table 6.)

Mpc	Light-year	km
1	3.26×10^6	3.09×10^{19}
3.07×10^{-7}	1	9.46×10^{12}
3.24×10^{-20}	1.06×10^{-13}	1

△ Table 6 Three units of distance: the megaparsec (Mpc), the light-year and the kilometre.

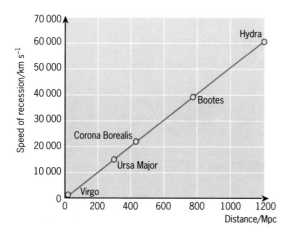

△ Fig 31 Graph of velocity of recession of galaxies plotted against distance.

QUESTIONS

1. State what the graph demonstrates.

2. Compare the times taken for the galaxies in Bootes and Corona Borealis to have arrived at their current locations after the Big Bang.

3. What is the gradient of the graph in km s⁻¹ Mpc⁻¹?

4. The gradient in km s⁻¹ Mpc⁻¹ is the Hubble constant, H. Assuming that the galaxies maintain their current recession velocities over time, how will H change over time? (You should consider how the gradient of the graph will change.)

5. Convert Mpc to km and recalculate the gradient and the unit.

6. For each galaxy a distance D away $v = D/T_H$ where T_H is known as the Hubble time or age of the Universe. Use this and the formula for the gradient of the graph to show that $T_H = 1/H$.

7. Calculate the age of the Universe in years.

8. Some well-established physical laws like the law of conservation of energy are considered to be principles. Why is Hubble's Law not considered to be a principle? [QWC]

> **Skills practised**
> 1, 12, 13, 14, 15, 16, 18, 26, 27

A2 ABSOLUTE ZERO

Just how cold can it get? Fahrenheit (1686–1736) worked with the practical assumption that the lowest attainable temperature was at the freezing point of salt water and set his (absolute) zero at that point.

The lowest natural air temperature recorded on Earth is −89.2 °C at Vostok, Antarctica. At this temperature carbon dioxide condenses from the atmosphere to form a solid carbon dioxide frost. This is considerably colder than 0 °F and 0 °C, so where is the absolute zero of temperature? In 1802 Gay-Lussac credited Jacques Charles with discovering in 1787 that all gases follow the same law of contraction when they are cooled. They contract by 1/273 of their volume for every degree Celsius fall in temperature. This strongly suggests that the intercept on the temperature axis at zero volume represents a lowest possible temperature – a true "absolute zero".

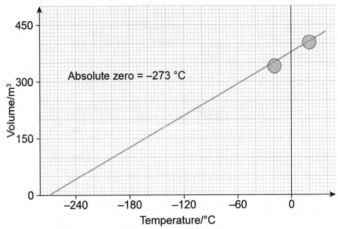

△ Fig 32 Volume against temperature in °C for a fixed quantity of an ideal gas.

The idea of a gas having zero volume at the absolute zero of temperature was explored by Lord Kelvin in 1848. The ideal gas law $PV = nRT$ shows that if the pressure on n moles of gas is maintained at some constant value then:

$$V = (nR/P) \times T$$

which predicts that $V \propto T$.

(R is the molar heat capacity or ideal gas constant, which is the amount of energy required to raise the temperature of one mole of a gas by 1 K.)

Using the triple point of water (0.01 °C) and degrees of the same size as Celsius degrees, the Kelvin scale of temperature begins at the absolute zero of temperature, equivalent to −273.15 °C.

That is, 0 K = −273.15 °C.

In 2003 a record low temperature of $(450 \pm 80) \times 10^{-12}$ K was reached. Scientists were investigating a state of matter known as a Bose–Einstein condensate, which only appears at very low temperatures. At the micro-scale, scientists use a different concept of temperature – one that links the kinetic energy of a gas atom or molecule to temperature,

$$\tfrac{1}{2}mv^2 = \frac{3kT}{2}$$

where m is mass, k is Boltzmann's constant and v is the root mean square (rms) speed of the gas particles. k is the

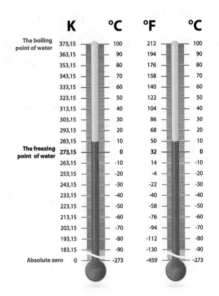

△ Fig 33 Celsius, Kelvin and Fahrenheit temperature scales.

specific molecular energy capacity, that is, the amount of energy required to raise the kinetic energy of one molecule by 1 K. Compare this to the meaning of the gas constant R.

There is therefore a clear relationship between the speed of a molecule and the temperature, $v \propto \sqrt{T}$. To double the speed of a molecule the temperature must increase four times.

$$R = 8.3145 \text{ J mol}^{-1}\text{K}^{-1}$$
$$k = 1.3806 \times 10^{-23} \text{ J K}^{-1}$$
$$v = \sqrt{3kT/m}$$

QUESTIONS

1. Taking 0 K to be –273 °C, what is room temperature 20 °C in kelvin?

2. What is the ratio of the volumes of one mole of air at 300 K to one mole of air at 100 K?

3. Show that the average relative molecular mass of air is 29 to two significant figures. Assume mass of air is 20% oxygen and 80% nitrogen.

4. Calculate the density of air at atmospheric pressure, 1.01×10^5 Pa and 0 °C. (*Hint*: Density = mass/volume. Calculate the mass and volume of 1 mole of air.)

5. What would be the mean speed of air particles at 1 K, if a gas state then existed?

6. From the answer to question 5 write down the root mean square speed of air particles at 400 K.

7. Calculate the energy of a single atom at 450×10^{-12} K in eV.

8. Classical thermodynamics is concerned with the macroscopic quantities of pressure, temperature and volume. Discuss the difficulties that arise when considering what happens when the temperature of a gas is reduced towards absolute zero. QWC

Skills practised

11, 12, 13, 14, 15, 16, 18, 25, 26, 27

A3 BLOODHOUND SSC

Bloodhound SSC is a supersonic car designed to reach 1000 mph (1600 km h^{-1}). All land speed record vehicles require large flat surfaces and that is why deserts are used. Like an aircraft, the vehicle will create a shockwave. At ground level this shock wave will interact with the desert floor, presenting in itself a considerable engineering challenge. Only one other car in history has broken the sound barrier – Thrust SSC in 1997.

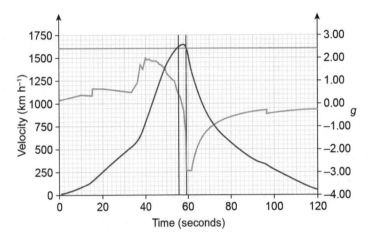

△ Fig 34 Velocity profile for Bloodhound SSC (courtesy Ron Ayers). Blue is velocity, red is acceleration and green is the speed of sound.

Exceeding the speed of sound at ground level will require enormous power. The car has two engines, a jet turbine (as used in the Eurojet fighter) and a rocket motor (Falcon Hybrid). The rocket will provide the force necessary to reach 1000 mph (1600 km h^{-1}) and has the advantage of not offering any air resistance as, unlike the jet, it has no air intake. Because it has no air intake the rocket is situated just behind the driver of the vehicle. The air intake for the jet engine is just above the driver's head!

Fig 34 shows the velocity (blue) profile and acceleration (red) profile for the car, with mass 7.8 tonnes fully fuelled. The jet afterburners are switched on 15 seconds from the start and the rocket is ignited at about 35 seconds from the start, providing maximum thrusts of 90 kN and 122 kN respectively.

Approximately 150 kg of jet fuel will have been burned at the 35 second mark. Without air resistance the acceleration of the vehicle driven by constant thrust would remain constant from 40 seconds to top speed and it would not be difficult to reach 1000 mph (1600 km h^{-1}). Note that during the jet only phase the acceleration slowly decreases and during the rocket and jet phase the acceleration decreases from the maximum of 2 g to zero at top speed. During this time air resistance is increasing as the square of the speed,

△ Fig 35 Bloodhound SSC.

which is why, despite the enormous thrust available, the forces are balanced at top speed where the acceleration is zero.

QUESTIONS

1. Estimate the length of the run from the area under the velocity–time graph.

2. What is air resistance at top speed?

3. Calculate the maximum power of the car in kW.

4. Using 746 W = 1 hp, calculate the power of Bloodhound SSC in hp. (hp or horsepower is still used to describe the power of Formula 1 racing cars. A Formula 1 car is about 700 hp.)

5. Verify, using the velocity–time graph, that the acceleration from 15 to 35 seconds is about 6 m s^{-2}, as shown by the acceleration–time graph.

6. At one stage in the journey, Bloodhound SSC covers 2.12 km from a starting velocity of 56 m s^{-1}. The acceleration is approximately uniform at 5 m s^{-2}. By setting up a quadratic equation using an equation of uniformly accelerated motion, calculate how long the stage of the journey took.

7. What is the net force on the car at 35 seconds before the rocket is ignited?

8. Estimate the air resistance at 35 seconds.

9. Just after the rocket has been ignited, at 40 seconds, the car has reached about half its maximum speed. If the thrust of the rocket at 40 seconds is 110 kN, show that air resistance will be approximately one quarter of the air resistance at top speed. State any assumptions you may make. **QWC**

Skills practised
2, 12, 13, 14, 15, 16, 18, 21, 23, 26, 28, 29

A4 DO WE NEED WIDE CAR TYRES?

The instinctive answer to this question is "Yes". In most people's minds the wide tyre is associated with better grip and an ability to brake in a shorter distance. Images of Formula 1 racing car tyres almost certainly reinforce the idea that wide tyres provide better stability and road handling, yet they do not!

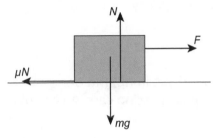

△ Fig 36 Recall the forces on a block at the point of sliding on a surface. Friction = μN where N is the force at right angles between the two surfaces.

Every country has a road use code and expects road users to learn about handling vehicles and the importance of keeping tyres in good condition. Drivers have to learn about braking distances and their dependence on bad weather. However, nobody has to learn about any differences in vehicles, because generally, given the same rubber compound and the same road surface, the braking distance for all vehicles is the same, whether it is a correctly laden 40-ton lorry or a small car or even a motorbike.

If stopping from a speed of v m s^{-1} the kinetic energy of the vehicle is transformed into heat via the braking system. The work done on the vehicle, mass m, is the frictional force between tyre and road, F_f multiplied by the displacement s from the point the brakes were applied.

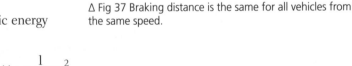

△ Fig 37 Braking distance is the same for all vehicles from the same speed.

Work done during braking = kinetic energy transformed into heat

$$F_f \times s = \frac{1}{2} mv^2$$

Also, $F_f \times s = \mu mgs$ (see Fig 36)

$s = v^2/2\mu g$ making s the subject of the equation

Braking distance s depends only on the initial speed and the nature of the surfaces, not on the mass of the vehicle.

The maximum frictional force available is directly proportional to the mass of the vehicle. It does not depend on the tyre tread, which is designed expressly for removing water on wet roads. The frictional force available is independent of the area of contact, meaning that narrow tyres or tyres without tread on a dry road will provide a proportionate frictional force to keep the braking distance the same (as the contact area decreases so the pressure exerted on the road increases). The width of tyres on Formula 1 cars is strictly controlled by regulations. The width of the tyre determines the rate at which it will wear out and so determines the number of pit stops. They are smooth for dry weather and treaded for wet. Making a misjudgment to use treaded tyres will mean the tyres wearing out more rapidly because of the greater pressure on the treaded portion and possibly an extra pit stop. High performance road cars in general have wide tyres only to stop them wearing out too quickly.

QUESTIONS

1. The minimum weight of a formula one car is 650 kg. Calculate its kinetic energy at 250 km h^{-1}.

2. How much heat is generated in the wheel brakes when coming to rest from 250 km h^{-1}?

3. What is the unit of the coefficient of friction?

4. If the coefficient of friction is 1.2, what is the minimum braking distance from 250 km h^{-1}?

5. Given the minimum braking distance from 250 km h^{-1} what is the maximum power dissipation in the brakes?

6. Why is there a minimum safe tread depth?

7. Why do wider tyres generally produce higher fuel consumption?

8. Discuss and explain how, if at all, wider tyres affect maximum cornering speed and braking distance. [QWC]

Skills practised
5, 11, 12, 13, 14, 15, 16, 18, 19, 26

A5 FLAT BATTERIES

Sources of potential difference, including cells, batteries and generators, have resistance, just like components such as resistors and bulbs. The resistance of sources is called internal resistance. These sources are either not connected to devices and are open circuit or they are connected and, closed circuit. Why do batteries fail so that on closed circuit they will not supply a current?

When a cell is open circuit, Fig 38(a), there is still a potential difference between its terminals even though it is not supplying a current. This potential difference, or open circuit voltage, is called electromotive force (emf), ε. (There is no good reason for calling this voltage emf, it is simply an accident of history.) With a modern digital voltmeter it is easy to measure ε as the resistance of these meters is so high (10–20 MΩ) they draw virtually no current.

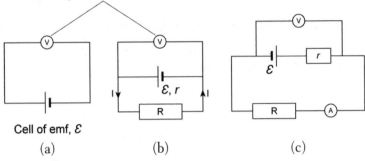

△ Fig 38 Cells showing (a) open circuit, (b) circuit closed across a resistor, and (c) a representation of internal resistance of the source, r.

When the circuit is closed across a resistor R, Fig 38(b), a current I is drawn and the voltmeter connected to the terminals of the cell will measure a lower voltage, the terminal or closed circuit voltage, V. The current I is through both the resistor and the **internal resistance** of the cell, r, so that:

$$\varepsilon = IR + Ir \quad \text{or} \quad \varepsilon = V + Ir$$

Rearranging gives:

$$V = \varepsilon - Ir$$

Ir is sometimes known as the "lost" volts.

△ Fig 39 Batteries.

QUESTIONS

1. Sketch a graph of V against I for the circuit (c), based on the equation $V = \varepsilon - Ir$

   a) What is the y-intercept?
   b) What is the x-intercept?
   c) What is the gradient?

2. A circuit is initially closed and consists of a 12 V battery, 3.7 Ω resistor and a switch, all in series. The internal resistance of the battery is 0.30 Ω. The switch is opened, so it then becomes an open circuit. What would a high resistance voltmeter read when placed:

   a) across the terminals of the battery?
   b) across the resistor?
   c) across the switch?

3. Repeat question 2 with the switch closed.

4. Torch batteries have an emf of about 1.5 V. Why could you not use eight of these batteries in series to start a car with a dead battery? (A car starter motor typically requires 100 amps at 12 volts to turn the engine over.)

5. High voltage (HV) power supplies used in schools have large internal resistances for safety. Why are they safer than HV supplies with a lower internal resistance? QWC

Skills practised
12, 13, 14, 15, 16, 18, 19, 26, 27

A6 SHOT PUT

This is a field event where the athlete must propel a mass of 7.26 kg (men) or 4.0 kg (women) and must ensure that the best possible angle for launch is chosen. As the shot moves freely under gravity after leaving the athlete's hand, it is a ballistic missile.

△ Fig 40 A shot putter in action.

Not only must the best angle be selected but the shot must leave the athlete's hand at as high a point as possible. This will ensure that the shot can travel for as long as possible and so cover the greatest possible horizontal distance, or range.

There is a small distance through which the shot can be accelerated before the athlete extends his arm. This distance is a result of the athlete's movement of his upper body. But this distance is limited by the circle in which the athlete can move.

The range R for a simple projectile landing at the same height from which it was launched is given by:

$$R = \frac{v^2 \sin 2\theta}{g}$$

where v is the launch velocity and θ is the angle of launch to the horizontal.

So, neglecting air resistance, the greatest horizontal range of a ballistic missile occurs when the launch angle is 45°.

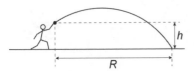

△ Fig 41 The flight of a shot.

The formula assumes that the vertical displacement at the end of the flight is zero when in fact the shot's final displacement is below the point from which the shot was released, h. So a more complete formula using h is given by:

$$R = \left(\frac{v^2 \sin 2\theta}{2g} \right) \left(1 + \sqrt{\left(1 + \frac{2gh}{v^2 \sin \theta} \right)} \right)$$

You can see that if h is zero the formula reverts to the first one for R. If h is not zero then the maximum range given by this formula is just a little less than 45° for given values of v and h.

However most shot putters launch at an angle much less than this. Only part of the energy supplied by the athlete goes into kinetic energy of the shot. The rest goes into raising the shot against gravity, and the greater the angle the more energy must be supplied and the lower the launch speed. Additionally the human arm develops less power the greater the angle. Both these limitations result in the launch angle for a world-class shot putter being anything from 26° to 38°.

QUESTIONS

1. Use the formula $R = \dfrac{v^2 \sin 2\theta}{g}$ and show why the maximum range occurs at 45°.

2. Set $v = 10$ m s^{-1} and $h = -2$ m and use a spreadsheet to model the fuller formula for R between 35° and 45°. From this determine the angle for the maximum range at this speed and height.

3. The horizontal and vertical motions for the simple projectile are treated separately so that

   $$h = y = v\sin\theta \times t - \frac{1}{2}gt^2 \qquad (1)$$

   $$x = v\cos\theta \times t \qquad (2)$$

   Set $h = y = 0$ and show that $R = \dfrac{v^2 \sin 2\theta}{g}$ by substituting from equation 1 into equation 2 (recall that $2\sin\theta\cos\theta = \sin 2\theta$).

4. If the men's shot is raised vertically 1.5 m, calculate the ratio of the change in gravitational potential energy, ΔGPE, to the change in its kinetic energy, ΔKE, on leaving the shot putter's hand at a speed of 8 m s^{-1}.

5. What is the horizontal velocity of the shot if it is launched at an angle of 35°?

6. Calculate the KE of the shot at the maximum height, neglecting air resistance.

7. Assuming the shot leaves the athlete's hand at 2.25 m above the ground, what is the maximum height reached?

8. Describe why the real launch height for putting the shot is around 35° rather than the 45° predicted by the range formula $R = \dfrac{v^2 \sin 2\theta}{g}$. **QWC**

Skills practised

3, 12, 13, 14, 15, 16, 17, 18, 19, 21, 22, 26, 28, 32

A7 BETA DECAY

In one form of radioactive decay a neutron decays into a fast electron (beta particle) and an anti-neutrino.

At the time nobody knew of the existence of the neutrino. What was known was that the spectrum of beta energies was very strange. Theory predicted that there would be a definite amount of total energy released in the beta decay. If two particles are involved in a disintegration then their momenta will be equal and opposite. The ratio of their kinetic energies will be the inverse of the ratio of their masses. So the beta particle should always take a fixed proportion of the total energy. The energy spectrum should be a sharp line at one particular energy.

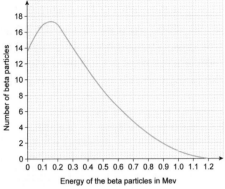

△ Fig 42 β-energy spectrum for bismuth-210.

This was not, however, what was seen. The energy spectrum for the beta particles was found to be continuous, with a well-defined maximum end point energy and a peak or mode at the lower energy end. This contradicted the principle of conservation of energy as nobody could account for the "missing" energy in the decays. Additionally, the recoiling nucleus and the beta particle did not always move in opposite directions.

In 1931 Pauli suggested that the end point energy was the total disintegration energy. He proposed that a third particle, the neutrino, shared the available kinetic energy with the beta particle. The neutrino would have very small mass and no charge and so would be virtually undetectable.

QUESTIONS

1. The equation for the decay of a neutron into a proton, electron and anti-neutrino is given by: $n^0 \rightarrow p^+ + e^- + \bar{\nu}_e$
   Show that mass number and electric charge are conserved.

2. Give the disintegration energy of bismuth-210 in eV.

3. What is the disintegration energy in joules?

4. Calculate the velocity of a beta particle if it takes all the available energy. (*Hint:* At relativistic speeds kinetic energy = $(\gamma - 1) m_0c^2$, where $\gamma = 1/\sqrt{(1 - v^2/c^2)}$ and m_0 is the rest mass of the electron, v is its velocity and c is the speed of light.)

5. Show that if two masses are involved in the decay their kinetic energy ratios are the inverse of their mass ratios.

6. Explain why the recoiling nucleus takes away little kinetic energy.

7. If the anti-neutrino takes away some kinetic energy, sketch a momentum diagram to show how momentum is conserved between all three particles.

8. In your own words, provide the argument to support the existence of the third particle in beta decay. [QWC]

Skills practised

1, 4, 11, 12, 13, 14, 15, 16, 18, 26, 30, 31

A8 THE WHEEL

The earliest records of wheels suggest that they appeared almost simultaneously in Europe and Mesopotamia about 5000 years ago. Wheels in museums show almost the same construction – a flat wooden disc with a hole for an axle.

Wheels rotating on moving vehicles have the curious property of always being stationary at one point – the point in contact with the ground.

To an observer on the ground a point on the rim of the wheel moves on a path known as a cycloid. To an observer on the hub of the wheel the same point appears to perform circular motion. The centripetal force which must act on a mass m in the rim is therefore given by:

$$F_c = mv^2/r$$

△ Fig 43 An early wooden wheel.

The Bloodhound supersonic car has solid wheels 0.90 m in diameter. The car itself will reach 1690 km h^{-1} and the acceleration at the rim of the wheel will be about 50 000 g. At these accelerations the strength of the material is of concern. Aluminium alloy will be used and this will have to be strong enough to provide the centripetal force needed at the rim where the acceleration will be 50 000 g. The maximum tensile strength of aluminium is 400 MPa.

$$strength = F_{max}/A$$
$$F_{max} = strength \times A$$

Will the strength of the material be sufficient to prevent it breaking up under high angular velocity? What follows is a very simplified approach to estimating whether or not a wheel rotating at very high speed will have structural integrity.

QUESTIONS

1. What is the circumference of the wheel?

2. Calculate the angular velocity of the wheel at top speed.

3. Show that $F_c = mr\omega^2$. (*Hint: $v = r\omega$ where ω is the angular velocity*)

4. Calculate the centripetal force needed on a 1 kg mass at the rim of the wheel.

5. If the density of aluminium is 2.7×10^3 kg m^{-3}, calculate the volume of a cube having a mass of 1 kg.

6. Calculate the cross-sectional area of the cube of mass 1 kg.

7. Calculate the force available to keep the mass of 1 kg at the rim rotating.

8. Compare your answers to questions 4 and 7 and discuss the problems facing the engineers in terms of the structural strength of aluminium and the integrity of the wheel. [QWC]

Skills practised
12, 13, 14, 15, 16, 18, 21, 26, 34

A9 THE ELECTROLYTIC CAPACITOR

At the heart of all timing circuits, switching circuits, ac filters, memory back-up and dc power supplies is the capacitor. Mistakenly believed to store charge, capacitors store electrical energy. They can release it rapidly, as in the camera flash-gun, or release it slowly through a large resistance, such as a memory card.

A capacitor is simply a pair of plates separated by a short distance. When the switch is closed or opened in a dc circuit as shown in Fig 45, the microammeter will flick briefly and then return to zero.

△ Fig 44 Capacitors on a sound card.

For a very short time charge has flowed onto one plate and an equal charge has flowed off the other plate. The resulting opposite charges on the two plates attract and hold each other in place through an electric field. With an equal and opposite charge on each plate the total charge is zero.

The plates now behave like the terminals of the battery but oppose any further flow of current. The current quickly reduces to zero when the capacitor is "fully charged". The potential difference across the plates of the capacitor is now equal and opposite to the potential difference at the terminals of the battery and Q is the charge on one plate.

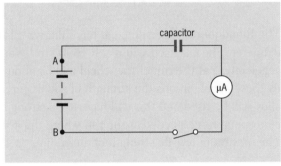

△ Fig 45 dc circuit with capacitor and microammeter.

Q is proportional to V and:

$$Q = C \times V$$

where C is the capacitance of the capacitor in Farads. (Note: do not confuse the C for coulombs with the C for capacitance.)

The energy stored in the capacitor is the work done to separate the charges. As the p.d. across the terminals increases, the charge separated increases. Since p.d. is joules per coulomb and $Q \propto V$, the area under the Q–V graph is the stored energy ($Q \times V \equiv C \times (J/C) \equiv J$).

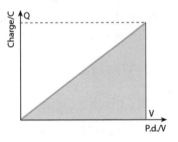

△ Fig 46 Charge against potential difference across the plates of a capacitor.

$$\text{stored energy } E = \Delta V \times Q = \frac{1}{2}QV$$

but $Q = CV$ therefore

$$\text{stored energy} = \frac{1}{2}CV \times V = \frac{1}{2}CV^2$$

The capacitance depends on the area of the plates and how close they are together. Both quantities will allow more charge to flow onto or off each plate.

$$C = \varepsilon A/d$$

where d is the distance between the plates and ε is a constant known as the permittivity of the space between the plates and measures the ease of setting up an electric field between them.

To make d as small as possible, an insulating aluminium oxide layer or dielectric is "grown" on one plate electrolytically, to create an "electrolytic" capacitor. This increases ε as well as decreasing d, but the molecules of the oxide layer are polarised so that the terminals of the electrolytic capacitor are labelled positive and negative to ensure correct connection. Connecting an electrolytic capacitor the wrong way around results in the molecules trying to reverse themselves and a lot of heat being generated.

Formerly capacitors were made by simply inserting an insulating layer such as polythene between two aluminium sheets so that it did not matter which terminal was connected to the positive or negative of the supply voltage. These capacitors could not be made small enough for use in modern appliances, which is why electrolytic capacitors have appeared. By controlling the "growth", the layer of oxide can be made very thin; it simply has the disadvantage of being polarised.

QUESTIONS

1. Show that the unit of ε is F m^{-1}.

2. The permittivity for an electrolytic capacitor is 6.4×10^{-11} F m^{-1}, the area of the plates forming the electrodes is 0.03 m^2 and the thickness of the dielectric is 10^{-8} m. Calculate the capacitance.

3. Describe what will happen if the capacitor is connected the wrong way around.

4. Calculate the energy stored in the capacitor at 200 V.

5. If this energy is released in 1/1000 s, calculate the power output.

6. Such a capacitor could be used for the flash-gun of a camera. Discuss the problems faced constructing a suitable capacitor for a small camera flash-gun. **QWC**

> **Skills practised**
> 11, 12, 13, 14, 15, 16, 18, 20, 26, 29

A10 PARTICLE OR WAVE?

In 1923 Louis de Broglie proposed that particles such as electrons should exhibit wave behaviour in much the same way as light had been shown to exhibit particle properties.

The story begins in 1901, with Planck suggesting that a solution to the problem of black body radiation was to restrict the energies of vibrations of atoms to multiples of their frequency of vibration multiplied by a small constant, which is now known as the Planck constant.

△ Fig 47 The French continue to recognise Louis de Broglie's groundbreaking work on this stamp.

$$E = 0, hf, 2hf, 3hf,..$$

This process is known as quantisation. The number h is very small, 6.63×10^{-34} J s. The energy hf is the energy quantum and the integer is the quantum number of the oscillator. The only justification that Planck could give for quantisation was that the mathematics worked and provided the observed form for the observed energy spectrum of a hot body (known as a black body). His mathematics agreed with the observation.

In 1905 Einstein took the idea further and showed that the photoelectric effect could be explained by quantising radiant energy. When radiant energy of sufficiently high frequency falls on a metal surface, electrons are ejected. These electrons are called photoelectrons and appear immediately when even the faintest radiation of the right frequency falls on the surface. The quanta of radiant energy are known as photons and behave like particles. For this discovery Einstein was awarded the Nobel Prize in 1921. The maximum kinetic energy of an ejected photoelectron is given by Einstein's photoelectric equation:

$$\frac{1}{2}mv^2 = hf - \Phi$$

Φ is known as the work function of the metal from which the electron is ejected and represents the attractive forces from the metal as the electron leaves the surface. Only electrons ejected at the surface will have the maximum kinetic energy. Quantisation of the radiant energy explains why photoelectrons can be ejected immediately radiation falls on the metal surface, as either the entire quantum is absorbed or not at all. As the frequency f of the radiation is reduced the maximum kinetic energy of the ejected electrons is reduced until it reaches zero. This minimum frequency is the threshold frequency.

De Broglie's hypothesis was to associate wave properties with particles using a combination of the energy quantum and $E = mc^2$. He proposed that the wavelength of a particle would be h/p where p is the particle's momentum. De Broglie was awarded the Nobel Prize for this work in 1929, two years after confirmation of the idea by Davisson and Germer, who showed that electrons were diffracted by nickel crystals. Both matter and radiation appear to exhibit both particle and wave properties!

QUESTIONS

1. What is the energy of a quantum of blue light in joules and electron volts?

2. Show that KE is related to momentum by $KE = p^2/2m$.

3. What is the de Broglie wavelength of an electron of 1 eV?

4. How does the wavelength of a particle change with velocity?

5. Show that the threshold frequency for ejected photoelectrons is given by $f = \Phi/h$

6. Φ for platinum is 9.9×10^{-19} J. Calculate the threshold frequency to eject photoelectrons from platinum.

7. Light of frequency 3.0×10^{15} Hz falls on platinum. Calculate the maximum velocity of the ejected photoelectrons.

8. Why should Davisson and Germer's experiment confirm de Broglie's hypothesis?

> **Skills practised**
> 1, 2, 3, 11, 12, 13, 14, 15, 16, 18, 26

A11 FUSION ENERGY

Average energy consumption worldwide is approximately 80 kWh per person per day. This huge appetite from seven billion people for energy demands that we seek more efficient ways to generate electrical energy. Electrical energy from nuclear fission is unpopular and energy from fusion is still in the experimental stage, largely at Cadarache in France where the International Thermonuclear Experimental Reactor (ITER) is being built. Commercial power production will be unlikely before 2050, after about 100 years of continuous research.

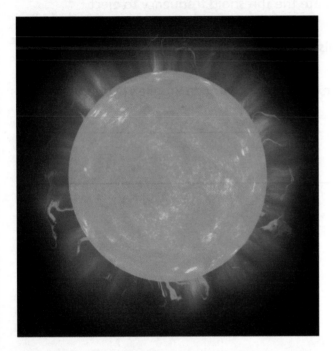

△ Fig 48 The Sun is a natural thermonuclear reactor.

The Sun is a natural thermonuclear reactor. Hot gas in the form of a plasma is contained by gravitational forces. At high temperatures particles move at greater speeds and when the temperature is high enough electrons are stripped from their nuclei so that electrons and nuclei move independently. This state of matter is a plasma. If the temperature is high enough the nuclei can collide with sufficient violence to fuse together. In the Sun the temperature is high enough for fusion reactions to occur and to create outward pressure great enough to stop the outer material of the Sun from collapsing inwards. Hydrogen nuclei or protons can fuse with each other to form helium-2, which immediately decays back into two protons! What is needed are neutrons to provide the nuclear force to hold the protons together. Very rarely at the point of fusion a proton can decay into a neutron, a positron and a neutrino via the same interaction responsible for β^- decay, the so-called weak interaction. This relatively rare event provides the necessary neutron so that the deuterium produced can fuse with another proton to form helium-3.

ITER is a tokomak device. A tokomak is a ring-shaped device or large torus (doughnut) in which the plasma is contained using magnetic fields. The plasma circulates in the middle and a large current through the plasma serves to generate a magnetic field that pinches and squeezes the plasma still further into the middle. This squeezing generates additional heat and the temperature is raised to about 150 million K. ITER is designed to operate at 50 MW input power generating 500 MW output power and will require about 100 kg of fuel a year (60 kg of deuterium and 40 kg of tritium). Compare this to 1.35 million tonnes of coal for a 500 MW coal-fired plant.

D	3.320×10^{-27} kg
T	4.970×10^{-27} kg
α	6.598×10^{-27} kg
n	1.675×10^{-27} kg
p	1.673×10^{-27} kg
β⁺	9.110×10^{-31} kg

The source of fusion energy is nuclear not chemical. Mass is not conserved during the fusion process and the "lost" mass is transformed into energy as predicted by the equation $E = mc^2$. Here on Earth we only intercept a very small amount of the Sun's radiation. The total solar output is equivalent to a mass loss of about 400 million tonnes per second.

QUESTIONS

1. Show that the total energy consumption worldwide per day is about 560 TWh.

2. How much energy is produced by a 500 MW generator in a year?

3. Assuming a continuous output at 500 MW, estimate the efficiency of a coal-fired power station if the heat of combustion of coal is 15 MJ kg⁻¹.

4. In the Sun, the net result from fusion is that four protons come together to form a helium nucleus and two positrons. Calculate the energy released from this net result. (*Hint:* First calculate the difference in masses between the initial particles and the final particles.)

5. In ITER, a deuterium nucleus D, fuses with tritium T to form a helium nucleus and a neutron. Calculate the energy released from this fusion reaction.

6. How many atoms of tritium are there in year's supply of energy for ITER?

7. If all the atoms of tritium are fused with deuterium, how much energy would be released in one year? How does this compare to your answer to question 2?

8. Discuss the problems of setting up a controlled nuclear fusion reaction. **QWC**

Skills practised

3, 11, 12, 13, 14, 15, 16, 18, 25, 26, 38

A12 ARTIFICIAL GRAVITY

In the film *2001 A Space Odyssey* an enormous space station is seen rotating in space as it orbits the Earth. This motion creates an artificial gravity but the direction of the artificial gravitational field is away from the central hub of the station. Inhabitants of the space station walk around the ring with their heads pointing towards the hub. A centripetal force acts on the feet of the inhabitants to push them in a circular path around the hub.

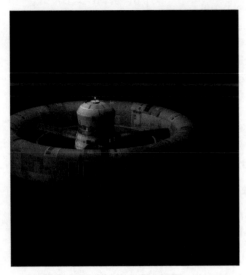

△ Fig 49 3D render of a spacestation.

A centripetal force is needed to move an object in a circular path. In a space station that force is the push from the outer wall towards the hub, so that inhabitants stand with their heads pointing towards the hub and their feet on the outer wall. As the wall pushes their feet so their feet push the wall, just as if they were in the gravitational field of the Earth.

The centripetal force is mv^2/r where m is the mass of the astronaut and r is the radius of the station. More massive people will therefore feel a bigger push against their feet and this will feel like a greater weight, just as on Earth. The tangential speed is v so that the angular velocity of the station is $\omega = v/r$. This means that increasing v will increase the size of the artificial gravitational field.

QUESTIONS

1. If the diameter of the station is 100 m what is the tangential speed necessary to create a field of 5 N kg^{-1}?

2. The station turns through an angle of 18.3° every second. What is this angle when expressed in radians?

3. Use $\omega = v/r$ to confirm the value for the angular velocity of the station given in question 2.

4. Engineers want to build another ring on the station so that the gravity field in the new ring will resemble the field on Earth, say 10 m s^{-2} or 10 N kg^{-1}. What will the radius of the new ring be if the angular velocity of the station is not changed?

△ Fig 50 A circus act from India.

5. A fairground activity known as the Rotor Ride or Gravitron consists of a large vertical wooden cylinder, 5 m in diameter. People stand against the inside of the cylinder, which is then rotated. At some point the floor is lowered and the riders remain "pinned" to the wall.

   a) What force prevents the riders from sliding to the bottom?
   b) The size of the frictional force between two surfaces is given by the coefficient of friction, μ, multiplied by the force at right angles to the surface, the normal force. For a person of mass m and a tangential velocity v, what is the size of the frictional force?
   c) Show that the tangential velocity needed does not depend on the mass of the riders.
   d) If the coefficient of friction is 0.1, at what tangential speed can the floor be lowered safely away without people sliding out?

6. Describe the motion of an astronaut "descending" an arm of the space station from the hub to the outer ring. **QWC**

Skills practised

6, 12, 13, 14, 15, 16, 18, 21, 26, 33, 34

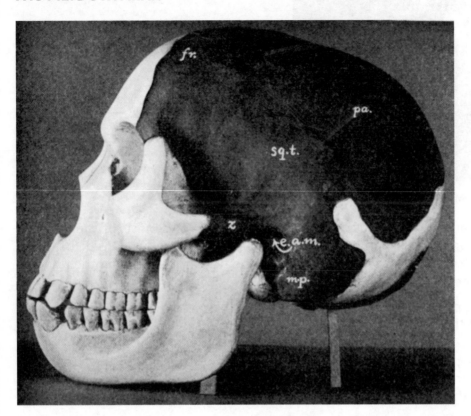

△ Fig 51 Model of skull of Piltdown man (*Eanothropus dawsoni*) as reconstructed by Dr Smith.

In 1912 part of the skull and jawbone of an early human was discovered in a quarry at Piltdown, England by Charles Dawson. He claimed the bones were half a million years old and were the missing link between apes and humans. The bones were controversial because some scientists interpreted the evidence differently, claiming a more modern origin. Finally in 1953, after more than 40 years, the bones were declared to be fake and much less than 50 000 years old. In 1959 radiocarbon dating revealed the skull and jaw to be about 500 years old, that the skull belonged to a human and the jawbone to an orangutan. It was the biggest hoax in paleontological history.

The method used to date the bones was to determine the amount of radioactive carbon-14 in the sample. When cosmic rays strike atoms in the upper atmosphere neutrons are sometimes ejected. These neutrons then strike nitrogen-14 nuclei and carbon-14 is produced. An equilibrium level of carbon-14 is reached so that the ratio of carbon-14 to carbon-12 in the atmosphere has remained constant over the millennia.

$$_{0}^{1}n + _{7}^{14}N \rightarrow _{6}^{14}C + _{1}^{1}p$$

As a natural material the bones would, when living, have exchanged carbon dioxide with the atmosphere and so maintained a constant ratio of carbon-14 to carbon-12. After death the ratio of radioactive carbon-14 to carbon-12 would decrease as the ^{14}C decays to ^{14}N.

$$_{6}^{14}C \rightarrow _{7}^{14}N + _{-1}^{0}\beta + _{0}^{0}\bar{\nu}$$

A beta particle, $_{-1}^{0}\beta$,

and an anti-neutrino, $_{0}^{0}\bar{\nu}$,

are produced. There are two ways of determining the amount of ^{14}C.

One is to measure the activity of the sample, counting the beta particles, and the other is to count the ^{14}C atoms present by burning the sample and converting the carbon into carbon dioxide. The gas is ionised and the ions are passed through a mass spectrometer so that the heavier carbon-14 ions can be separated out and counted by measuring the ionisation current. This second method is particularly useful when the activity is likely to be very small. The ratio of carbon-14 to carbon-12 in living things is about $1.5:10^{12}$, so is already very small.

QUESTIONS

1. Show that both nuclear transformations above conserve mass number and charge.

2. The half-life of ^{14}C is 5700 years. How would the activity of an 11 400-year-old wooden object compare to the activity of the same object made recently?

3. Write down an equation connecting the number of radioactive atoms, N, present in a sample t years old to the number of radioactive atoms, N_0, in the sample when it was fresh.

4. The average ionisation current at the carbon-14 detector in the mass spectrometer is 10 pA for 1.6 seconds. Assuming singly ionised molecules arrive, calculate the number, N, of carbon-14 atoms present.

5. Using $^{14}C:^{12}C = 1.5 : 10^{12}$ in living material calculate the number of carbon-14 atoms, N_0, expected in a reference sample if the number of carbon-12 atoms is 8.0×10^{19}.

6. Estimate the age t, of the sample.

7. Comment on the assumptions made and the problems associated with using carbon dating, in this case and in general. QWC

> **Skills practised**
> 12, 13, 14, 15, 16, 17, 18, 26, 29, 35, 38

A14 TIME LORDS

Simple harmonic oscillators are real Time Lords! The pendulum, balance wheel, quartz crystal and caesium atoms are all simple harmonic oscillators and are all used to measure the passage of time.

The key to understanding why simple harmonic oscillators are good time-keepers is simply to know that their vibrations are isochronous, meaning that the periods of their vibrations are equal even as the oscillations die away through damping. When a pendulum swings, the period for small oscillations is the same as that for larger oscillations. The greater the frequency of the oscillator the better it is for keeping time as each single oscillation is a shorter time and therefore errors will be smaller. Caesium atoms vibrate at 9 192 631 770 Hz, which is considerably larger than the one or two Hz of a pendulum.

Analysis shows that if the acceleration, a, of an oscillator has the form $-ky$ where y is the displacement from the equilibrium position, then the oscillations will be simple harmonic. Further analysis shows that $k = \omega^2$ so that $a = F/m = -\omega^2 y$ for a simple harmonic oscillator.

Imagine a rod of circular cross-section floating vertically in a liquid of density ρ. If the rod is lifted a little and then released it will bob up and down.

Will the oscillations be simple harmonic? If so, what is the period of the oscillations?

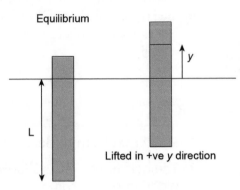

△ Fig 52 Rod floating vertically in liquid.

QUESTIONS

1. The density of the liquid is ρ and the mass of the rod with a uniform cross-sectional area A is m.

   a) What is the mass of the displaced liquid in the equilibrium position in terms of A, L and ρ?

   b) Write down an equation connecting the two forces acting on the rod in the equilibrium position.

   c) Taking the upwards direction as positive, show that the acceleration of the rod when lifted a little from the equilibrium position and released will be given by

   $$ma = -yA\rho g$$

2. Now that you have shown that $a = -(A\rho g/m)y$, which is the requirement for simple harmonic oscillations, write down an equation for ω^2 in terms of A, ρ, g and m and then solve for ω.
3. Since $\omega = 2\pi f$, what is the period T in terms of ω?
4. Substitute for ω and show that $T = 2\pi\sqrt{m/(A\rho g)}$
5. What is the period of oscillation of a buoy, mass 1000 kg, diameter 1 m, in fresh water?
6. Discuss how you could use a bobbing stick to measure the density of a liquid.

> **Skills practised**
> 5, 12, 13, 14, 15, 16, 18, 21, 26, 30, 34

A15 TO THE MOON

Jules Verne's great story *From the Earth to the Moon*, written in 1865, was finally realised in 1968 when astronauts Lovell, Borman and Anders orbited the Moon in Apollo 8 and returned to Earth. Verne knew about the inverse square law of gravitational attraction and notes that the "weight of the projectile will decrease rapidly, and will end up by being completely annulled at the moment when the attraction of the Moon will be equal to that of the Earth".

Verne's projectile Columbiad was launched from a cannon 300 metres long and had to reach a speed of 11 000 m s^{-1} if it was to reach the Moon. Columbiad would take 83 hours to reach the point where the Moon's and Earth's gravitational force were equal and opposite (the neutral point) and a further 14 hours to reach the Moon, 97 hours in total. Apollo 8 took 69 hours to reach the Moon and 53 hours to return to Earth. (Verne's distance to the neutral point was 47/52 of the Earth–Moon distance.)

To escape from Earth a projectile needs sufficient kinetic energy (KE) to be able to slow down indefinitely, transforming KE to GPE but never actually stopping. At a sufficiently great distance, therefore, its KE will be zero and so will its gravitational potential energy (GPE) as it will not fall back to Earth. The only way for GPE to increase to the value zero is for the GPE to be negative everywhere. Since potentials are scalar quantities (remember energy is a scalar) the presence of the Moon near Earth has the effect of lowering the potential everywhere as the two negative potentials when added simply become more negative, that is, they decrease. This means that as the projectile, all the time slowing down, reaches the point between Earth and the Moon

△ Fig 53 Earthrise from Apollo 8.

where the gravitational force is zero, the potential energy will begin to decrease towards the Moon and the projectile will then accelerate towards it.

Gravitational potential, V, simply means potential energy per unit mass or potential energy per kilogram.

$$V = \frac{Gm_E}{r}$$

where m_E is the mass of Earth. The potential energy of mass m kg a distance r from Earth is then Gm_Em/r joules. The gravitational force on m kg is the familiar:

$$mg = \frac{Gm_Em}{r^2}$$

QUESTIONS

1. Calculate escape velocity from Earth (ignore the presence of the Moon). (Earth's mass = 5.97×10^{24} kg and radius = 6.37×10^6 m.)

2. Calculate the ratio of Earth to neutral point distance to the Moon to neutral point distance. (Moon's mass = 7.35×10^{22} kg).

3. How does your answer to question 2 compare to Verne's calculation?

4. What is the distance from Earth to the neutral point if the mean distance to the Moon is 3.84×10^8 m?

5. What would be the average g force experienced by Columbiad during the launch from the cannon? (Take 1 g to be equivalent to 10 m s^{-2}.)

6. Why is there a difference between the time it took Columbiad to reach the Moon and the time it took Apollo 8?

7. Explain how Apollo 8 was able to reach the Moon with an initial velocity less than the escape velocity from Earth. QWC

Skills practised

6, 12, 13, 14, 15, 16, 17, 18, 24, 26, 28

AIS1

When placing a ladder against a wall it is important to ensure that the angle the ladder makes with the ground is big enough to provide a frictional force at the base that is sufficiently large to stop the ladder from slipping as a worker climbs to the top.

Fig 54 is a free body diagram with all the forces acting on the ladder shown. It has been assumed that the ladder is light (has no mass) and that the wall is smooth (no friction between ladder and wall).

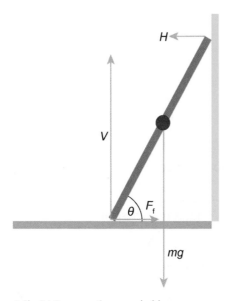

△ Fig 54 Forces acting on a ladder.

A worker of mass m begins to climb the ladder and reaches the point marked by the blue circle.

Vertically: $V = mg$ (the forces are balanced vertically).

Horizontally: $F_f = H$ (the forces are balanced horizontally).

The length of the ladder is L and the distance along the ladder to the worker from the bottom is d.

As the worker climbs the ladder, H and F_f will increase in size. This is because the worker's weight and the push from the wall both have moments about the bottom of the ladder and any turning effects of forces must also balance if the ladder is not to slip.

Moments about the bottom of the ladder: $H \times L\sin\theta = mg \times d\cos\theta$ **(equation 1)**

But $F_f = H$ therefore equation 1 becomes:

$$H = F_f = mg \times \frac{d}{(L \times \tan\theta)}$$ **(equation 1a)**

If the coefficient of friction between the ground and the base of the ladder is μ, then the frictional force is: $F_f = \mu \times V = \mu \times mg$ **(equation 2)**

Therefore equating the right-hand sides of equations 1a and 2 and making $\tan\theta$ the subject of the equation gives: $\tan\theta = d / \mu L$

SECTION A

A student decides to carry out an experiment to measure μ, the coefficient of friction for a situation similar to that described above. She models the real-life set-up with a model ladder 1 m long. She increases the value of d and finds the value of d at which the ladder just slips. At this point she measures the angle the ladder makes with the ground using a protractor.

The results are shown in Table 7.

d/cm ± 0.2 cm	θ	$\tan\theta$
50.0	44	
60.0	50	
70.0	56	
80.0	59	
90.0	66	

△ Table 7

1. What is the independent variable in this experiment? **(1)**

2. Show that $\tan\theta = d/\mu L$ using equations 1a and 2 as described above **(max 3)**

3. The student wishes to plot the data and obtain a straight line plot. She compares the equation from question 2 to $y = mx + c$. What is the physical quantity represented by the gradient? **(1)**

4. State two sources of systematic error in carrying out this experiment. **(2)**

5. Estimate the percentage errors in θ and d for the largest value of d. **(2)**

SECTION B

In order to gain an accurate value for μ the student decides to plot a graph and determine the gradient. She chooses to calculate $\tan\theta$ rather than measuring θ with a protractor. To do this she measures the Y distance up the wall and the X distance along the floor when the ladder is just about to slip for each value of d (see Table 8).

d/cm ± 0.2 cm	Y/cm ± 0.2 cm	X/cm ± 0.2 cm	$\tan\theta$
50.0	72.4	69.0	
60.0	79.2	61.0	
70.0	84.8	53.0	
80.0	87.7	48.0	
90.0	90.7	42.1	

△ Table 8

1. Estimate the percentage error in $\tan\theta$ using this method. **(1)**

2. Why do you think the student chose to calculate $\tan\theta$ rather than using a protractor to measure θ directly? You may need to consider the uncertainty in each method. **(max 2)**

3. Complete Table 8 with values of $\tan\theta$. **(2)**

4. Draw a graph of $\tan\theta$ against d. **(3)**

5. Calculate the gradient of the line. **(3)**

6. Calculate μ, the coefficient of friction between the ground and the base of the ladder, if the length of the ladder $L = 1.000 \pm 0.002$ m. **(2)**

7. Estimate the percentage error in μ using $d = 90.0$ cm from the table of results. **(3)**

8. Estimate the absolute error in μ. **(2)**

9. The experiment was not very repeatable and uncertainties in Y and X were closer to 1 cm than 0.2 cm. How many significant figures would be sensible for μ? **(2)**

10. An organisation for health and safety recommends an angle θ of 75° with the ground for placing a ladder against a wall. Why? **(3)**

Skills practised

7, 8, 9, 10, 20, 25

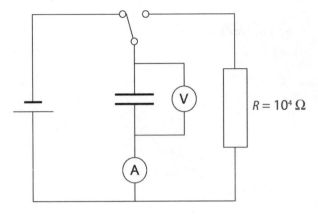

△ Fig 55 The discharging resistor, $R = 10^4\ \Omega$

SECTION A

A student explored the charging and discharging of a capacitor using a datalogger. The capacitance C was measured by plotting a discharge curve of V against t. The relation between V and t ($V = V_0 e^{-t/RC}$) was used to find C. An exponential curve was fitted to the data.

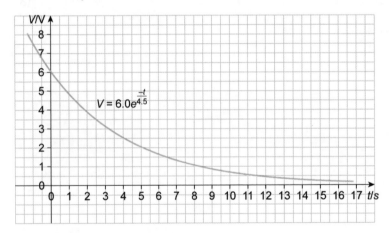

△ Fig 56 Exponential curve based on capacitance.

1. What is the initial value of V? (1)

2. What are the units of $R \times C$? (1)

3. What is the value of RC? (2)

4. When $t = RC$ what is the value of V? (1)

5. Calculate the value of C. (2)

SECTION B

In this second part, a student without a datalogger has a capacitor of unknown size and a resistor of 10 kΩ. He decides to measure the capacitance by discharging the capacitor through the resistor and measuring the potential difference (p.d.) across the capacitor using a high-resistance voltmeter.

The method chosen is to measure the initial pd, V_0 across the capacitor when fully charged, then to discharge the capacitor through the known resistor and to start timing immediately the switch is thrown. The voltmeter is watched until it reaches 1 volt and the time, t_1, noted. This is repeated a number of times to take an average of V_0 and t_1.

Theory shows that $\ln V = \ln V_0 - t/RC$.

This is a straight-line relationship between $\ln V$ and t. The student now has the data (see Table 9) to mark the intercept on the $\ln V$-axis and the intercept on the t-axis. A straight line can be drawn and the gradient measured. From the gradient, C can be calculated.

	$V_0 \pm 0.1/V$	$t_1 \pm 0.2/s$
	6.1	8.2
	6.1	8.1
	6.0	8.0
	6.1	8.2
	5.9	7.9
Average		

△ Table 9

1. With the switch as shown in Fig 55, what will the ammeter reading be after a short time? **(1)**

2. Sketch a graph of V against t and on the same axes I against t for the discharging of the capacitor through the resistor. Your sketch should show the initial values of V and I, and the values of t and I when $V = 1$. **(4)**

3. What are the percentage uncertainties in the average values of V_0 and t_1? **(2)**

4. What is the physical quantity represented by the gradient of the graph of $\ln V$ against t? **(1)**

5. What is the theoretical fractional change in V_0 after RC seconds? **(1)**

6. Find the average of V_0 and t_1. Then draw a graph of $\ln V$ against t using just the two points ((average t_1), 0) and (0, $\ln$(average V_0)) that you can calculate from the table. **(3)**

7. Sketch on the same axes a graph of $\ln V$ against t when the charging voltage V_0 is just 3.0 V. **(2)**

8. Measure the gradient of the graph in question 6. **(3)**

9. Use the gradient to calculate C. **(2)**

10. Estimate the percentage uncertainty in the gradient from your answer to question 3. **(2)**

11. Estimate the absolute uncertainty in C, given no error in R. **(2)**

12. From your graph, what is V after RC seconds? **(1)**

13. Compare your answer to the fractional change in V_0 from question 12 with the fractional change in V_0 from question 5 and discuss the method of finding C using just two experimental points. **(2)**

14. A laboratory decides to determine the value of the capacitance using this method of timing until $V = 1\,V$. To improve the technique they use a range of different resistors.

R/Ω	V/V	t_1/s	$\ln V$	C/F
1.00×10^6	6.0	790		
1.00×10^5	6.1	80.0		
1.00×10^4	6.0	7.95		
1.00×10^3	6.1	0.81		

△ Table 10

Complete Table 10 and estimate a mean value for C including the uncertainty from the range. **(3)**

Skills practised
7, 8, 9, 10, 20, 25, 35, 36, 37

Answers

Each activity's set of questions gives you the opportunity to practise and develop skills covered in the **Skills** section and, by answering them, you will enhance your understanding of the skills necessary for success on your course.

This section includes all the answers, as well as helpful hints and tips to boost your performance in that exam room. In most cases the working has been provided. To conserve space the answers have often been written linearly with something = something = something. An attempt has been made to ensure that the equals signs are logically consistent. An example of an illogical set of equals statements would be:

work done = 15 J = power = 15/3 = 5 W

Clearly 15 ≠ 15/3 and this sort of writing should be avoided. Where space is not an issue, equations should always be rewritten underneath each other with the equals signs lined up. For example:

work done = 15 J

    power = work done /time taken

        = 15/3 W

    power = 5 W

Practise writing out the necessary data neatly and ensure that all parts of an equation have been included. Take care with denominators and numerators and what parts should be calculated first. Tidiness is very important in keeping a track of what has been done when calculating a large formula. There are plenty of questions for you to practise number crunching in these activities.

Although the Estimating skill (S38) is more about being able to arrive at a quick answer to a calculation and deals with the powers of 10 separately, remember also that you might be asked to estimate quantities, that is to write down a sensible value. For example, the mass of a human being would be about 70 kg, but anything from 60–80 kg would suffice.

QWC The last question in each activity is a Quality of Written Communication (QWC) question. These test your subject knowledge and understanding and ask you to think about how you communicate your ideas. For these questions, we have not only outlined the subject knowledge points you need to cover, but have also included an indication of what low, medium and high scoring responses would do to show you how to improve. For more guidance on QWC questions, we have written low, medium and high scoring responses to Activity 5, Question 5 in the **QWC Worked Examples** section. Each response has detailed points commenting on its quality of written communication, giving you an insight into what's needed.

The **Assessing Investigative Skills** section gives you the chance to develop your investigative skills in two major activities without the need to complete practical work. You will be able to tackle AIS1 soon after you've started your course and AIS2 will help you tackle a more advanced investigation later on. The answers to these offer comprehensive guidance so you can check your responses and see how to progress.

Formulae and data are of little use unless you develop the skills to connect them together. The activities are wide ranging and should test your ability to draw on subject knowledge gained throughout your course. Always remember that critical attention to detail is what drives physicists and it is this that will ultimately help you to improve your scores.

A1 Age of the Universe

1. The graph shows that the speed at which galaxies are receding appears to be increasing with increasing distance from the Earth.

2. Same time because everything started at the same place and time.

3. Gradient = 60 000/1200 = 50 km s^{-1} Mpc^{-1}

4. At later times the galaxies will be at greater distances. If the velocities remain the same then the gradient will decrease with time and so H will decrease with time. (Today H is believed to be about 70 km s^{-1} Mp s^{-1}, giving an age of the Universe of about 14 billion years.)

5. Gradient = 50/(3.09 × 10^{19}) = 1.6 × 10^{-18} km s^{-1} km^{-1} = 1.6 × 10^{-18} s^{-1}

6. $v = D/T_H$ so $T_H = D/v$. But $v = HD$ from the graph, so $T_H = 1/H$ by substitution

7. $T_H = 1/1.6 × 10^{-18} = 0.625 × 10^{18}$ s
   $T_H = 0.625 × 10^{18} / 3.15 × 10^7 = 2.0 × 10^{10}$ years

Note that 3.15 × 10^7 is the number of seconds in one year.

8. **QWC** Your answer should include the following points:

   - Hubble's Law is based on empirical evidence.

   - It simply states the relationship between two variables whereas principles usually provide an explanation.

   - It should be possible to use theory to make predictions that agree with Hubble's Law.

   - Scientists are still using different interpretations to explain the evidence.

   - It is not possible to test Hubble's Law in controlled conditions, e.g. in the laboratory. This makes it harder to study and understand.

   - A principle is theory so well established that all scientists are in agreement.

   - For example, the law of conservation of energy is a principle. (Other examples, such as momentum, would also be acceptable.)

A low level answer would include one of these points.

A medium level answer would include two, three or four points and contain more detail, such as a discussion of the uncertainty regarding the theory behind the observations rendering it less well understood.

A high level answer would include five or six points.

A2 Absolute zero

1. 293 K

2. $V_{100}/100 = V_{300}/300$ therefore $V_{100}/V_{300} = 100/300 = 1/3$
   $V_{300} : V_{100} = 3 : 1$

3. Assuming 20% oxygen and 80% nitrogen,
   average relative molecular mass
   $= [(80 \times 28) + (20 \times 32)] /100 = 29$

4. $V = nRT/P$

   From the formula above the volume of one mole of air can be calculated. The mass of one mole of air is needed to calculate the density. Since one mole of air contains an Avogadro's number of molecules, the mass of one mole can be calculated using the average relative molecular mass from question 3.

   If $n = 1$ mole, then,

   Volume of 1 mole $= nRT/P = 1 \times 8.3 \times 273/1.01 \times 10^5$
   $= 2.24 \times 10^{-2}$ m^3 (a fact well known to chemists).

   Mass of 1 mole $= 29 \times 10^{-3}$ kg

   Density $= m/V = 29 \times 10^{-3}/2.24 \times 10^{-2}$
   $= 1.3$ kg m^{-3}

5. $v = \sqrt{(3kT/m)}$
   $= \sqrt{[3 \times 1.4 \times 10^{-23} \times 1/(29 \times 10^{-3}/6.0 \times 10^{23})]}$
   $= 29$ m s^{-1}

6. At 400 K, $v = 29 \times \sqrt{400} = 580$ m s^{-1}

7. Joules into eV divide by 1.6×10^{-19}
   $E = [(3 \times 1.38 \times 10^{-23} \times 450 \times 10^{-12}) / 2] / 1.6 \times 10^{-19}$
   eV $= 5.8 \times 10^{-14}$ eV

8. **QWC** Possible marking points:
   - Classically the volume of an ideal gas at absolute zero is zero.
   - Classically the pressure of an ideal gas at absolute zero is zero.
   - Real gases are composed of molecules.
   - The molecules themselves have a fixed volume.
   - Therefore, long before absolute zero is reached the molecules would find themselves with no space in which to vibrate, but as the temperature is above absolute zero they must be vibrating.
   - In practice a real gas would solidify or liquefy long before absolute zero.
   - Solidification or liquefaction fixes the volume above zero.

A low level answer would include one or two of these points, probably the first two given here.

A medium level answer would include three or four points giving more detail, such as a discussion of the uncertainty regarding the theory behind the observations rendering it less well understood.

A high level answer would include five or six points and include some discussion of what happens to real gases as they are cooled.

A3 Bloodhound SSC

1. Select rectangles of size 250 km s^{-1} by 20 s. There are 12.5 of them.

   Area under v–t graph = displacement
   $= (250 \times 1000/3600) \times 20 \times 12.5$

   Length of run = 17 km

2. At top speed air resistance = max. thrust
   $= 90 + 122 = 212$ kN

3. $P = Fv = 122 \times 10^3 \times (1690 \times 1000/3600) = 1.0 \times 10^8$ W

4. $P = 10^8/ 746 = 134\,000 = 1.3 \times 10^5$ hp

5. Gradient at 35 = 300 km h^{-1}/(35 − 20) = 20 or about 6 m s^{-2}. Red line showing acceleration is approximately horizontal at 0.6g and g = 9.8 m s^{-2} giving a = 5.9 m s^{-2}.

6. $s = 2120$ m $\qquad u = 56$ m s^{-1} $\qquad a = 5$ m s^{-2}
   $s = ut + \frac{1}{2}at^2 \qquad 2120 = 56t + 2.5t^2$
   $2.5t^2 + 56t − 2120 = 0$

   $$x = \frac{-b \pm \sqrt{b^2 - 4ac}}{2a}$$

   $$= \frac{-56 \pm \sqrt{56^2 - 4 \times 2.5 \times (-2120)}}{2 \times 2.5}$$

   $= 20$ s or $−42.4$ s

   The only sensible answer is 20 s.

7. Net force $= ma = 7.8 \times 1000 \times 6 = 4.6 \times 10^4$ N

8. Air resistance 90 kN − 46 kN = 44 kN

9. **QWC** Your answer should include the following points:
   - At the 40-second mark, the maximum acceleration is about 20 m s^{-2}.
   - The net force is given by ma, which is $(7.6 \times 1000 \times 20)$ or 150 kN.
   - The jet fuel has been assumed depleted by 200 kg.
   - The total thrust from the engines at the 40-second mark is 200 kN assuming the jet is operating at full thrust.
   - Therefore the net force = 200 − air resistance = 150 kN.
   - Air resistance is approximately 50 kN, which is about one-quarter the air resistance at top speed.

- Air resistance is proportional to v^2.
- Doubling the speed should increase the air resistance four times.

A low level answer would include one or two of these points.

A medium level answer would include three or four points.

A high level answer would include five or six of the stages above, with working clearly shown.

A4 Do we need wide car tyres?

1. $0.5 \times 650 \times (250 \times 1000/3600)^2 = 1.6 \times 10^6$ J

2. Approximately 1.6×10^6 J

3. $\mu = F_f / N$ or the ratio of two forces so μ has no unit.

4. $F_f \times s = 1.2 \times 650 \times 9.8 \times s = 1.6 \times 10^6$ J

   $s = 200$ m

5. $a = -u^2/2s$ so $a = -12.1$ m s^{-2}

   Time taken to stop $= (v - u)/a = -69.4/-12.1 = 5.74$ s.

   Power dissipation $= 1.6 \times 10^6 / 5.74 = 2.8 \times 10^5$ W or 280 kW.

6. This is the depth at which the tyre can still safely remove water from the road.

7. Wheels are more difficult to accelerate being more massive.

8. **QWC** Your answer should include the following points:
   - Frictional force provides the centripetal force necessary for cornering.
   - Friction will not depend on the width of the tyre.
   - Friction does not depend on area of contact.
   - Friction $= \mu N$.
   - $N = mg$.
   - A heavier vehicle has a greater weight and therefore greater friction.
   - The braking distance will not depend on mass or type of vehicle.
   - The frictional force depends only on the coefficient of friction between the tyre and road.
   - A wet road reduces μ for all vehicles and increases stopping distance.
   - A wet road reduces μ for all vehicles and lowers maximum cornering speed.

A low level answer would include one or two of these points, including at least one of the statements regarding the independence of tyre width and stopping distance or cornering speed.

A medium level answer would include three or four points giving more detail, such as one of the two equations stated.

A high level answer would include five or six points including a chain of reasoning to explain why stopping distance and/or braking distance is unaffected by tyre width.

A5 Flat batteries

1. *Hint*: scales on axes are not required as the question only asked you to sketch the graph.

   a) y-intercept is ε volts, the open circuit voltage when $I = 0$.

   b) x-intercept is known as the short circuit current, ε/r amps.

   c) Gradient is $-r$ ohms, the internal resistance of the cell.

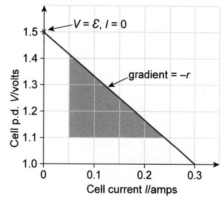

△ Fig 57 Answer to question 1.

   You should have been able to sketch this by looking at

   $V = \varepsilon - Ir$

   This is the graph of a straight line, y-intercept ε, x-intercept ε/r and gradient $-r$. In practice the short circuit current is not something you want to measure! The short circuit current is measured by connecting the terminals of the battery together. It is the x-intercept and is the largest current that can be drawn from the battery.

2. In the first case the switch is open, no current flows, this is open circuit.

   a) 12 V

   b) 0 V

   c) 12 V (Take care here as students often think the voltage across the switch is 0 V all the time.)

3. With the switch closed:

   $I = 12/(3.7 + 0.3) = 3$ A. Note that is necessary to calculate the current flowing before calculating the terminal voltage.

   a) $V = 12 - 3 \times 0.3 = 11.1$ V

   b) 11.1 V, same as voltage across battery, connections are in same place.

   c) 0 V

4. Torch batteries are <u>likely to</u> have a much higher internal resistance than a car battery. You simply cannot draw a large current from them. The short circuit current of a dry cell is probably only about 1 A and the current needed is about 100 A. The internal resistance of a car battery is so low (maybe only a milliohm) that it can easily deliver such a large current without a noticeable drop in its terminal voltage.

5. **QWC** The answer to this question is in the QWC Worked Examples section.

A6 Shot put

1. When $\theta = 45°$, $2\theta = 90°$ and so $\sin 2\theta = 1$, which is its maximum value. So for any v the maximum range occurs when $\theta = 45°$.

2. If your table is correct you should get 12 m at about 40°.

3. From 1: $v\sin\theta \times t = \frac{1}{2}gt^2$ therefore $t = 2v\sin\theta/g$

   Substituting into 2: $x = v\cos\theta \times t$
   $= v\cos\theta \times 2v\sin\theta/g = v^2\sin 2\theta/g$

4. ΔGPE/ΔKE = $7.26 \times g \times 1.5/0.5 \times 7.26 \times 8^2 = 107/232$

   Dividing top and bottom by 107 gives
   ΔGPE/ΔKE = 1 : 2.2

5. $v\cos\theta = 8 \times \cos 35° = 6.6$ m s^{-1}

6. $0.5 \times 7.26 \times 6.6^2 = 158$ J, which is the kinetic energy at the highest point.

   At maximum height GPE gained = KE lost
   $= 232 - 158 = 72$ J

7. $7.26 \times g \times h = 72$
   $h = 72/(g \times 7.26) + 2.25 = 1.01 + 2.25 = 3.26$ m

8. **QWC** Your answer should include the following points:

   - The landing point is lower than the launch point.
   - The force an athlete exerts is different at different angles, due to the structure of the human body.
   - Different athletes have different builds, so each will have a different optimum angle.
   - They tend to exert large forces at shallower angles.
   - A lot of energy goes into just lifting the shot to the launch height…
   - …which involves increasing its gravitational potential energy.
   - It may better to lower the launch height and so reduce the launch angle to allow the shot to gain more kinetic energy/speed.

   A low level answer would include one or two of these points, including that different people have different body shapes or the force that can be applied differs with angle and the shot lands lower than its launch point.

   A medium level answer would include three or four points including those in a low level answer but also with some mention of energy considerations.

A high level answer would include five or six points, including an appreciation of less GPE allowing more KE to be imparted to the shot.

A7 Beta decay

1. Mass number = 1 to left and right as e⁻ and the anti-neutrino both have mass number 0 and proton and neutron both have mass number 1. Electric charge to the right = 1 − 1 = 0 = charge to the left.

2. 1.2×10^6 eV

3. $1.2 \times 10^6 \times 1.6 \times 10^{-19} = 1.9 \times 10^{-13}$ J

4. If you were unaware that beta particles were emitted at relativistic speeds you would write $\frac{1}{2}mv^2 = 1.9 \times 10^{-13}$ and find $v = 6.5 \times 10^8$. This is well in excess of the speed of light. You might then have suggested that the speed of the beta particle must be very close to the speed of light and earn at least one mark. You should use the relativistic formula for calculating the kinetic energy, KE.

   $KE = (\gamma - 1)m_0c^2$, where $\gamma = 1/\sqrt{(1 - v^2/c^2)}$.
   $\gamma = [(1.9 \times 10^{-13})/(9.1 \times 10^{-31} \times 9.0 \times 10^{16})] + 1 = 3.32$
   $\sqrt{(1 - v^2/c^2)} = 1/3.32$ $v^2/c^2 = 1 - (1/3.32)^2 = 0.909$
   $v = 0.95c = 2.8 \times 10^8$ m s^{-1}

   This was not an easy question. You have to know that beta particles move at relativistic speeds and then the solution is in two steps, first γ then v.

5. $p_1 = p_2$ $m_1v_1 = m_2v_2$ $(m_1/m_2)^2 = (v_2/v_1)^2$
   $m_1/m_2 = m_2v_2^2/m_1v_1^2 = KE_2/KE_1$

6. It is relatively very massive, so from question 5 the bigger the mass, the smaller the KE.

7. See Fig 58.

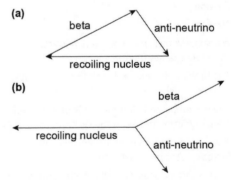

Fig 58 Answer to question 7.

8. **QWC** Your answer should include the following points:

   - If different nuclei of the same isotope decay, the kinetic energy available should be the same each time.
   - If just two particles are involved in the disintegration they have equal and opposite momenta (momentums).

- All beta particles will have the same kinetic energy, for the same isotope decay.
- However, beta energies range from small to a well-defined maximum.
- Energy cannot be destroyed; it must be conserved.
- There must be a third particle carrying away the rest of the energy when the beta's energy is below the maximum.
- The energy of the beta particle and the neutrino will add up to the well-defined maximum amount.
- The energy is randomly shared between the beta particle and the neutrino.
- Very little kinetic energy is taken away by the recoil nucleus, even though its momentum might be large, because its mass is relatively much larger than the beta or the anti-neutrino.

A low level answer would include one or two points, probably stating that the existence of the neutrino is necessary for conservation of energy.

A medium level answer would include three or four points, including more detail about why a neutrino is needed to conserve energy and how they would all be expected to have the same energy in its absence.

A high level answer would include five or six points, including a good level of detail as to what happens to the energy available after a decay.

A8 The wheel

1. $2\pi r = 2 \times 3.14 \times 0.45 = 2.83$ m
2. 1690 km h^{-1} $= 1690 \times 1000/3600 = 469$ m s^{-1}

   $\omega = 2\pi f$ and f is found by dividing 2.83 into 469 to find the revolutions/second.

   $f = 469/2.83 = 166$ Hz

   $\omega = 1040$ rad s^{-1}
3. $F_c = mv^2/r = mr^2\omega^2/r = mr\omega^2$
4. $F_c = 1 \times 0.45 \times 1040^2 = 4.87 \times 10^5$ N
5. Volume $= \frac{1}{2} . 7 \times 10^3 = 3.7 \times 10^{-4}$ m^3
6. Area of cross-section $= (0.37 \times 10^{-3})^{2/3}$ (that is, the cube root of the volume, then squared)

   $= 0.37^{2/3} \times 10^{-2} = 5.2 \times 10^{-3}$ m^2
7. $F_{max} = 400 \times 10^6 \times 5.2 \times 10^{-3} = 2.08 \times 10^6$ N
8. **QWC** Your answer should include the following points:
   - $F_{max} > F_c$
   - F_{max} is about 20 times larger than F_c
   - So the wheel will probably not disintegrate at these very high rotational speeds.
   - This gives a good safety margin (engineers typically work on a safety factor of four times). Bloodhound appears safe.

- However, if the wheel hits rocks/bumps in the road …
- … damage/cracks may occur, …
- which could weaken the structure.
- Friction could cause the wheel to get very hot.
- Materials become softer at higher temperatures.
- So the melting point of the aluminium needs to be much higher than its running temperature.

A low level answer would include one or two of these points, including the realisation that $F_{max} > F_c$ and may include the conclusion Bloodhound is safe.

A medium level answer would include three or four points, including the idea the $F_{max} \gg F_c$ and therefore there is a good safety margin involved.

A high level answer would include five or six points, including the ratio of F_{max} to F_c and a discussion of the need to take potential damage *or* heating into account.

A9 The electrolytic capacitor

1. $C = \varepsilon A/d$, $\varepsilon = Cd/A$
   Replacing the quantities with their units:
   units of ε: Fm/m^2 = Fm^{-1}
2. $C = (6.4 \times 10^{-11} \times 0.030)/1 \times 10^{-8} = 0.19 \times 10^{-3}$
   $= 2 \times 10^{-4}$ F or 200 µF
3. Heat will be generated, the capacitor will get hot and could explode. The dielectric layer will be destroyed, rendering the capacitor useless.
4. $E = \frac{1}{2} CV^2 = \frac{1}{2} \times 200 \times 10^{-6} \times 200^2 = 4$ J
5. $P = E/t = 4/10^{-3} = 4$ kW
6. **QWC** Your answer should include the following points:
   - The capacitor will need to be small.
   - Being small may reduce the energy storage capacity/the power output because:
     - the area of the plates has to be minimised
     - the dielectric as has to be as thin as possible.
   - Using a larger p.d. would store more charge but:
     - the capacitor could break down (develop a hole in the dielectric)
     - the p.d. is limited to a small value by the batteries powering the camera.
   - A thin dielectric is liable to puncturing and makes the capacitor fragile, which renders it useless.

A low level answer would include one or two of these points, probably including the conflict between the need for a small size and storing enough charge.

A medium level answer would include three or four points, including those in a low level answer plus some mention of the practicalities of making a small capacitor store enough charge.

A high level answer would include five or six points including those in a medium level answer plus the consequences of making the capacitor smaller.

A10 Particle or wave?

1. $E = hf = 6.6 \times 10^{-34} \times (3.0 \times 10^8 / 0.4 \times 10^{-6})$ taking the wavelength of blue light as being 400 nm. (The value used here is 400 nm, which some sources give as violet light.)

    $E = 5.0 \times 10^{-19}$ J

    $E = 5.0 \times 10^{-19}/1.6 \times 10^{-19}$ eV $= 3.1$ eV

2. KE $= \frac{1}{2} mv^2 = \frac{1}{2} mv^2 \times m/m = \frac{1}{2} (mv)^2/m = p^2/2m$

3. $\lambda = h/p = 6.6 \times 10^{-34}/\sqrt{(2 \times 9.1 \times 10^{-31} \times 1.6 \times 10^{-19})}$
    $= 1.2 \times 10^{-9}$ m or 1.2 nm

4. If m is constant $\lambda \propto 1/v$. (λ is inversely proportional to v.)

5. At the threshold $\frac{1}{2} mv^2 = 0$ so $\Phi = hf_t$

    $f_t = \Phi/h$

6. $f = 9.9 \times 10^{-19} / 6.6 \times 10^{-34} = 1.5 \times 10^{15}$ Hz

7. $\frac{1}{2} mv^2 = hf - \Phi = hf - hf_t = h(f - f_t)$

    $\frac{1}{2} mv^2 = 6.6 \times 10^{-34} \times (3.0 - 1.5) \times 10^{15} = 9.9 \times 10^{-19}$ Hz

    $v^2 = 2 \times 9.9\ 10^{-19}/ 9.1 \times 10^{-31} = 2.18 \times 10^{12}$

    $v = 1.5 \times 10^6$ m s^{-1}

8. **QWC** Your answer should include the following points:

    - Diffraction is a property of a wave motion.
    - The electrons reflecting off the crystal showed diffraction effects.
    - This confirmed the idea that travelling electrons behaved as a wave motion.
    - The crystal was behaving like a diffraction grating.
    - The electrons were behaving like waves being diffracted.
    - A pattern similar to the diffraction pattern was seen when light passed through a gap was detected.
    - The diffraction pattern was pronounced because the spacing between layers of atoms in the crystal was similar in magnitude to the wavelength of the electrons.
    - Diffraction of a wave is most pronounced when the gap/obstacle is about the same magnitude as its wavelength.
    - The greater the momentum of a particle the shorter its wavelength, so the smaller the gap/object needed to diffract it significantly.

    A low level answer would include one or two of these points, probably including the idea that diffraction is a wave property leading to the conclusion that electrons had wave properties.

A medium level answer would include three or four points, possibly including the idea that the planes of atoms within the crystal acted as a diffraction grating.

A high level answer would include five or six points, including making the link between a higher momentum and shorter wavelength.

A11 Fusion energy

1. 80 kWh $\times 7 \times 10^9$ people $= 560 \times 10^9$ kWh
    $= 560 \times 10^{12}$ Wh $= 560$ TWh

2. $500 \times 10^6 \times 365 \times 24 \times 3600 = 1.6 \times 10^{16}$ J

3. $(1.6 \times 10^{16})/(1.35 \times 10^6 \times 10^3 \times 15 \times 10^6)$
    $= 1.6 \times 10^{16}/ 20.3 \times 10^{15}$

    Efficiency = 79%, which is rather optimistic as no power station can operate at maximum power continuously!

4. $4 \times 1.673 \times 10^{-27} = 6.692 \times 10^{-27}$ kg = mass 4 protons

    α + 2 positrons $= 6.600 \times 10^{-27}$ kg

    $\Delta m = 0.092 \times 10^{-27}$ kg. $E = \Delta mc^2 = 8.28 \times 10^{-12}$ J

5. $\Delta m = (8.280 + 8.273 - 6.598 - 1.675) \times 10^{-27}$ kg
    $= 0.017 \times 10^{-27}$ kg. $E = 1.53 \times 10^{-12}$ J

6. Number atoms $= (40\ 000/3) \times 6 \times 10^{23}$
    $(N = (M/m) \times N_A)$

    number $= 8 \times 10^{27}$

7. Energy released $= 8 \times 10^{27} \times 1.53 \times 10^{-12} = 1.2 \times 10^{16}$ J. This compares well with question 2. The answers are similar but you would expect the energy from the fuel to be greater than 1.6×10^{16} J. Even nuclear powered generators do not run at 500 MW continuously, so the energy output will be much less than 1.6×10^{16} J, as for the coal-fired station.

8. **QWC** Your answer should include the following points:

    - The very hot plasma (1.5×10^8 K) must be kept away from the walls of the tokomak.
    - Hot plasma would vaporise any material it encountered/It cannot be contained by matter.
    - The ions are moving at very high speeds/KEs.
    - The ions require very strong magnetic field to contain them.
    - If the plasma squeezes through the magnetic field, high-speed computers will be needed to adjust the field shape very rapidly.
    - As the neutrons are neutral they will escape the plasma and hit the walls of the tokomak.
    - The walls have to be cooled to remove the thermal energy from neutrons.
    - A 50 MW power station will be needed to operate the tokomak.

A low level answer would include one or two of these points, possibly related to the first two here.

A medium level answer would include three or four points and contain more detail, such as why it is necessary to contain the plasma with a magnetic field.

A high level answer would include five or six of these points.

A12 Artificial gravity

1. $v^2/r = 5$ therefore $v = \sqrt{(5 \times 50)} = 16$ m s^{-1}

2. Angle in radians $= 18.3 \times 2\pi/360 = 0.32$ rad

3. $\omega = v/r = 16/50 = 0.32$ rad s^{-1}

4. $v^2/r = r^2\omega^2/r = 10$

   $r = 10/\omega^2 = 10/0.32^2 = 98$ m

5. a) Friction

   b) mg or better $\mu mv^2/2.5$

   c) $\mu mv^2/r = mg$ $v = \sqrt{(rg/\mu)}$ so v does not depend on m.

   d) $v = \sqrt{(2.5 \times 9.8/0.1)} = 16$ m s^{-1}

7. **QWC** Your answer should include the following points:

   - Centripetal acceleration supplies an acceleration equivalent to g.
   - The lesser/greater the distance from the centre a ring is the lower/higher the speed.
   - The lower/higher the speed the lesser/greater the acceleration/g.
   - At the hub acceleration/$g = 0$.
   - The maximum value of acceleration/g occurs at the outermost ring.
   - $a = \omega^2 r$.
   - As the astronaut descends an arm the side walls will push him/her tangentially.
   - As there is nothing under the astronaut's feet to push them into a circular path the astronaut will begin to accelerate down the arm.
   - It will feel as though there is a force away from the hub towards the outer rings.
   - Because $a = \omega^2 r$, as radius increases so does acceleration because angular speed is constant.
   - The centripetal/resultant force is caused by the normal reaction force applied to the astronaut by a ring.

A low level answer would include one or two of these points, probably stating one of the extremes of the artificial gravitational field at the hub or maximum distance for the centre.

A medium level answer would include three or four points including more detail, such as a discussion of the speed increasing as the radius increases.

A high level answer would include five or six points, including an understanding that it is specifically the normal reaction force a ring applies on the astronaut that supplies the resultant force **and** that this increases for rings further from the centre.

A very good high level answer would include the medium level points and also mention one of the last two points here.

A13 Piltdown Man

1. Eq 1: $1 + 14 = 14 + 1$ so mass number is conserved

   $0 + 7 = 6 + 1$ so charge is conserved

   Eq 2: $14 = 14 + 0 + 0$ so mass number is conserved

   $6 = 7 + (-1) + 0$ so charge is conserved

2. After two half-lives the activity would be a quarter of the original.

3. $N = N_0 e^{-\lambda t}$

4. $Q = I \times t$

   $N = Q/(1.6 \times 10^{-19}) = (I \times t)/(1.6 \times 10^{-19})$
   $= 10 \times 10^{-12} \times 1.6/(1.6 \times 10^{-19}) = 1.0 \times 10^8$

5. Number of ^{14}C: Number of ^{12}C $= 1.5 : 10^{12}$

   $N_0/8.0 \times 10^{19} = 1.5/10^{12}$ therefore
   $N_0 = 1.5 \times 8.0 \times 10^{19}/10^{12} = 1.2 \times 10^8$

6. From question 3, $1.0 \times 10^8 = 1.2 \times 10^8 e^{-\lambda t}$

   $\ln 1.2 = \lambda t = [(\ln 2)/T_{1/2}]t$

   $t = 0.182 \times 5700/0.693 = 1500$ years to two figures.

7. Your answer should include the following points:

   - The rate of production of carbon-14 has been assumed constant over time.
   - Possible reasons why not:
     - In recent years nuclear weapons testing has raised the levels of carbon-14 in the atmosphere
     - Cosmic ray bombardment may not have been constant over thousands of years.
   - However, comparisons of carbon dating with other methods (e.g. tree ring dating) agree well.
   - Contamination of small samples is always a problem despite efforts to ensure cleanliness.
   - Only once-living things can be dated, i.e. a flint axe head cannot be dated but a wooden axe handle could be.
   - Manufactured tools could have been repaired over time with newer material.
   - It gets harder to date very old objects (many times the half-life of carbon-14) because the amount of carbon-14 left becomes very difficult to measure accurately (you cannot date dinosaur bones using this method, for example).

A low level answer would include one or two of these points.

A medium level answer would include three or four points, including a possible factor that could affect the accuracy of the measurements.

A high level answer would include five or six points including that dating becomes more challenging as the age of the sample increases or it is not possible for very old samples.

A14 Time Lords

1. a) $m = AL\rho$

   b) $mg = AL\rho g$

   c) Net force $= ma = A\rho(L - y)g - mg$
      $= AL\rho g - Ay\rho g - mg = AL\rho g - Ay\rho g - AL\rho g$
      $ma = -yA\rho g$

2. $\omega^2 = A\rho g/m$ so $\omega = \sqrt{(A\rho g/m)}$

3. $\omega = 2\pi f = 2\pi/T$ $T = 2\pi/\omega$

4. $T = 2\pi/\sqrt{(A\rho g/m)} = 2\pi\sqrt{(m/A\rho g)}$

5. $T = 2\pi \sqrt{1/(\pi \times 0.25 \times 9.8)} = 2.26$ s

6. **QWC** Your answer should include the following points:

   - The stick has to float vertically…

   - …so at least half of it has to be submerged/its centre of mass has to be submerged at all times.

   - It has to have a lower density than the liquid it is displacing.

   - It has to be more than half the density of the liquid (if its mass is uniformly distributed).

   - The higher/lower the density of the liquid the higher/lower it sits in the liquid.

   - The bobbing time period could be measured…

   - …by calculating an average from a total of at least 30 periods summed from repeat reading.

   - Density can be found by using the equation from question 6.

   - Rearranged, the equation from question 6 is $\rho = (4\pi^2 m)/(T^2 Ag)$.

   A low level answer would include one or two of these points probably taken from among the first five and discuss the qualitative ideas about the height the stick sits in the water compared with the density of the fluid.

   A medium level answer would include three or four points including discussing the need to measure the time period (although not necessarily from more than one value for T) and use of the equation – the rearrangement may not be shown.

   A high level answer would include five or six points including discussing how it is necessary to measure the mean time period from repeated readings and use of the rearranged equation.

A15 To the Moon

1. $(\tfrac{1}{2}) mv^2 = (Gm_E \times m)/r_E$

   $v = \sqrt{[(2 \times 6.67 \times 10^{-11} \times 5.97 \times 10^{24})/(6.37 \times 10^6)]}$
   $= 11\,200$ m s^{-1} or 1.12×10^4 m s^{-1}

2. At the point where $Gm_E/r_E^2 = Gm_M/r_M^2$, where the suffixes E and M stand for Earth and the Moon respectively, the forces of attraction from Earth and the Moon are balanced. The distance to the point from Earth is r_E and the distance from the Moon is r_M.

   Cancelling G and taking the square root of both sides $\sqrt{(m_E/m_M)} = r_E/r_M$

   $\sqrt{(5.97 \times 10^{24}/7.35 \times 10^{22})} = 9.0$

   so $r_E / r_M = 9 : 1$

3. 47/52 (from the text) gives $r_E/r_M = 47/(52 - 47)$
   $= 9.4 : 1$.

4. Distance $= (9/10) \times 3.84 \times 10^8 = 3.46 \times 10^8$ m

5. $a = (v^2 - u^2)/2s = 11\,000^2/(2 \times 300) = 2.0 \times 10^5$ m s^{-2}
   or 2.0×10^4 g.

6. As Verne's neutral point was bigger than the fraction of the Earth–Moon distance that we know it to be, the Columbiad would have spent longer slowing down in Earth's gravitational field.

7. **QWC** Your answer should include the following points:

   - Escape velocity means the minimum velocity with which a missile must be fired from Earth, without ever falling back.

   - Normally the craft would eventually fall back to Earth at less than escape velocity.

   - KE lost as GPE gained.

   - The presence of the Moon lowers the potential everywhere.

   - The craft can arrive as far as the neutral point with less than escape velocity.

   - It must have some kinetic energy at the neutral point (but it can be close to zero).

   - Motion then continues towards the Moon at neutral point

   - The craft will fall towards the Moon from the neutral point.

   - The craft will still not be able to escape the Earth–Moon system.

   - As the potentials of both Earth and the Moon are negative everywhere, each lowers the potential due to the other at all points in their combined field.

   - Use of a diagram, such as Fig 59, to show the lowering of the potential by the Moon.

   A low level answer would include one or two of these points, probably showing an appreciation of GPE increasing as the craft moves away from Earth.

A medium level answer would include two or three points including appreciating the commensurate fall in KE as GPE increases and how this is reversed past the neutral point.

A high level answer would include four or five points and would show an understanding that the Moon lowers the potential so that the craft does not need as much KE to "escape" to the Moon as it does to escape Earth completely.

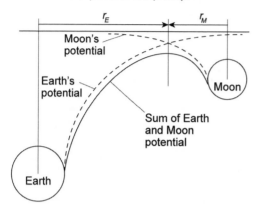

△ Fig 59 The Moon lowers the potential everywhere (question 7).

AIS1

Section A

1. d

2. $H \times L \sin\theta = mg \times d \cos\theta$

   $\sin\theta/\cos\theta = \tan\theta = mgd/HL$
   $(V/F_r) \times (d/L) = d/\mu L$ (one mark for each equality above, up to a maximum of 3)

3. Gradient $= 1/\mu L$ m^{-1}

4. The ladder has weight (will increase the gradient) and the wall will have some friction (will decrease the gradient). Marks for identification only.

5. Error in θ: $(1/66) \times 100 \approx 1.5\%$
   Error in d: $(0.2/90) \times 100 \approx 0.2\%$

Section B

1. Adding the percentage errors in X and Y, error in $\tan\theta = 0.7\%$.

2. Error in angle using a protractor will not be better than 1°, suggesting % error at say 60° is $[(\tan 60 - \tan 59)/\tan 60] \times 100 \approx 4\%$, greater than 0.7%.
   Or 2 marks for estimating the % error in 1° in 60° say 2%.

3. 1.05, 1.30, 1.60, 1.83, 2.15 (at least 3 correct to no more than four significant figures, 1 mark; all five correct 2 marks).

4. Sensible selection of scales (1 mark). Correct plotting (1 mark) and line drawn (1 mark).

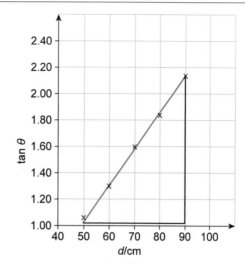

△ Fig 60 Answer to question 4.

5. Gradient $= \Delta\tan\theta/\Delta d$ where $\Delta\tan\theta$ and Δd are as large as possible (1 mark). Gradient $= (2.15 - 1.03)/0.400 = 2.80$ m^{-1}, 1 mark for correct value to at least 3 significant figures, no more than 4 significant figures.

6. $1/\mu L = 2.80$ (1 mark), $\mu = 1.0/2.80 = 0.357$ (1 mark).

7. % error in $d = (0.2/90) \times 100 \approx 0.2\%$
   Total percentage error in gradient $= 0.7 + 0.2 = 0.9\%$ (1 mark).
   % error in L is $(2/1000) \times 100 = 0.2\%$ (1 mark).
   Total error in μ is 1.1% (1 mark).

8. 1.1% of $0.357 = 0.0041$ (1 mark), $\mu = 0.357 \pm 0.004$ (1 mark).

9. With five times the error in $\tan\theta$, the total error will be almost five times greater (1 mark), suggesting just two significant figures for μ (1 mark), $\mu = 0.36 \pm 0.02$ (1 mark also possible here but no more than 2 marks for the entire question).

10. Tan 75° $= 3.73$ (1 mark). In the experiment the worker could easily climb to the top of the ladder with $\tan\theta$ having a maximum of about 2.5 (1 mark), so any angle bigger than 68° would be possible. But too large an angle could result in the ladder falling backwards away from the wall if the centre of gravity falls on the wrong side of the base (1 mark).

AIS2

Section A

1. 6 V

2. seconds

3. 4.5 (1 mark) seconds (1 mark)

4. $V = 6/e = 2.21$ V or about 2.2 to 2.3 V from the graph.

5. $C = 4.5/R$ (1 mark) $= 4.5 \times 10^{-4}$ F (1 mark, including the unit).

Section B

1. 0 A

2. See Fig 61. 1 mark for $I_o = 0.68$ mA. $I = 0.10$ mA when $V = 1$, 1 mark. The point (8.1, 1.0), 1 mark. I having same shape as V, 1 mark.

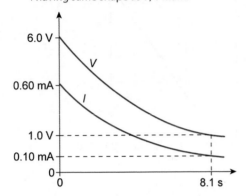

△ Fig 61 Answer to question 2.

3. $(0.1/6.0) \times 100 \approx 2\%$ error in V_0
   $(0.2/8.1) \times 100 \approx 2.5\%$ error in t_1

4. $-1/RC$ in s^{-1}

5. When $t = RC$, $\ln V = \ln V_0 - 1$, $\ln (V_0/V) = 1$, $V_0/V = e$. Therefore $V = V_0/e$ or the fractional change in V_0 is about 0.37.

6. Averages 6.0 V and 8.1 s to two significant figures. If you have used three significant figures then your answers to questions 7, 8 and 9 will change slightly but it makes no difference to your answer to question 11. See Fig 62. 1 mark for 1.79, 1 mark for t-intercept, 8.1. 1 mark for drawing straight line.

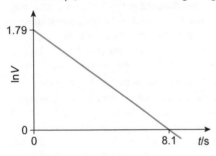

△ Fig 62 Answer to question 6.

7. One mark for $\ln V = 1.10$ or 1.1 and one mark for parallel or cutting at 4.9 or 4.94. See Fig 63.

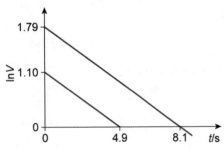

△ Fig 63 Answer to question 7.

8. Gradient $= -(1.79/8.1) = -0.221$ s^{-1}. (1 mark for 3 figures using V_o and t-intercept, 1 mark for correct answer and 1 mark for unit).

9. $-1/RC = -0.221$ (1 mark). $C = 1/(10^4 \times 0.221)$ $= 4.52 \times 10^{-4}$ F (1 mark including unit).

10. % error in gradient = % error in $\ln V_0$ + % error in t_1 (1 mark) = 2 + 2.5 = 4.5% (1 mark)

11. 4.5% of 4.52 = 0.2 (1 mark); $C = 4.5 \pm 0.2 \times 10^{-4}$ F. (1 mark including unit).

12. $RC = 4.5 \times 10^{-4} \times 1.0 \times 10^4 = 4.5$ s
    $\ln V = 0.80$, $V = 2.23$ V

13. 2.23/6.0 = 0.37. Therefore to two significant figures the fractional decrease in voltage is what would be expected (1 mark and 1 mark for any of the following statements). Since the straight line must go through the intercepts on the axes the method of repeating to find the best possible values of the intercepts appears valid. There will be sources of systematic error in the measuring instruments for voltage and time along with any systematic delay in noting the time when the voltmeter reads 1 V. The value of R of the known resistor will also introduce some error.

14. See Table 11.

R/Ω	V/V	t_1/s	$\ln V$	C/F
1.00×10^6	6.00	790	1.792	4.41×10^{-4}
1.00×10^5	6.10	80.0	1.808	4.42×10^{-4}
1.00×10^4	6.00	7.95	1.792	4.44×10^{-4}
1.00×10^3	6.10	0.81	1.808	4.48×10^{-4}

△ Table 11 Answer to question 14.

The range is 0.07×10^{-4} F suggesting the uncertainty is 0.04×10^{-4} F.

Mean $C = (4.44 \pm 0.04) \times 10^{-4}$ F. (1 mark for completing the table and one each for C and the uncertainty).

QWC Worked Examples

At some point during your course, you will be assessed on the quality of your written communication. These annotated worked examples show how low, medium and high scoring responses to Question 5 in Activity 5 gain the marks they do.

5. High voltage (HV) power supplies used in schools have large internal resistances for safety. Why are they safer than HV supplies with a lower internal resistance?

In general write short sentences. Answer the question carefully. Pay careful attention to any command words. Question 5 uses the command word "why" and specifically asks for a comparison between the HV supply with high internal resistance and the HV supply with low internal resistance. The "why" requires some reasoning. You will only get one mark for simply stating that the "school HV supplies only give low currents". Do not provide irrelevant information.

The five skills listed for quality of written communication can best be remembered as:

1 Think of your intended audience.

2 Use short sentences.

3 Organise your thoughts clearly.

4 Be accurate with technical language.

5 Look for command words.

★ Low level answer

The information conveyed by the answer is poorly organised and may not be relevant or coherent. There is little correct use of specialist vocabulary. The form and style of writing is only partly appropriate.

The expression "not as good as others" is not very technical and is popular language.

Sloppy spelling, punctuation and grammatical errors make this answer difficult to understand.

It is not clear from the answer whether or not the student understands that the high internal resistance of the school HV supply is a safety feature or is simply related to poor quality.

There is limited use of Physics terminology as the only terms used, such as "internal resitances", are in the original question.

The student knows that, in general, power supplies have low internal resistances but does not explain why this makes them safer.

> *The school HV supply is not as good as others. power suplies should have low internal resitances otherwise they do not work well. Thats' why they are safe.*

★★ Medium level answer

The information conveyed by the answer is not well organised or fully coherent. There is little use of specialist vocabulary, or specialist vocabulary may have been used incorrectly. The form and style of writing is not really appropriate.

There is some technical language such as "current will flow" but not a sufficient amount.

Spelling and grammar is good in this example.

There is no attempt to compare the two supplies.

The expression "to keep you safe" is not very scientific and would be regarded as popular language.

There is some attempt to provide some reasoning and therefore tackle the "explain" in the original question. The third sentence may well be correct but is not relevant to the question.

> *The higher internal resistance will protect you. If you connect yourself to the terminals only a small current will flow through you because there is a high resistance. School power supplies have fuses to keep you safe.*

★★★ High level answer

The information conveyed by the answer is clearly organised, logical and coherent, using appropriate specialist vocabulary correctly. The form and style of writing is appropriate for answering this type of question.

The student has avoided any slang or popular terminology, using correct Physics terminology written in a formal style instead.

The spelling and grammar are very good, especially as the answer includes a number of technical terms all spelled correctly.

The student has compared the two supplies and stated why one is safe and the other is dangerous.

The expression "low internal resistance" indicates some technical mastery of the topic.

The student has explained why the internal resistance causes a terminal voltage drop and that low voltages are relatively safe.

If the power supply has a low internal resistance, it can deliver a large current. Anyone connected across the terminals of a HV supply with a low internal resistance will get a high voltage shock, suffering possible electrocution, as the terminal voltage will not change much. If the internal resistance is high, the terminal voltage will immediately drop when attempting to draw a large current. The value of the internal resistance is chosen to bring the terminal voltage down to a safe value. Normally such supplies can only deliver currents of a few milliamps and this is stated on the front of the casing. Any attempt to draw a larger current simply lowers the terminal voltage.

Glossary

arithmetic This applies to the scale used in graphs. Arithmetic scales are those where the scale markers simply add up, for example 3, 6, 9, 12, etc. In an arithmetic progression the sequence of numbers is such that the difference between consecutive terms is constant. The difference in the example provided is 3.

asymptotic Approaching but not quite getting there. In geometry an asymptote of a curve is a straight line such that the distance between the line and the curve approaches zero as they tend to infinity. It is not always possible to draw an asymptote to a curve; for example, it is not possible to draw an asymptote to a parabola. The asymptotes to $y = 1/x$ are the x and y axes.

base The base of a number system is the number of different unique digits including zero used to represent numbers. Base 10 has 10 different digits and there is a clear connection between powers and bases. For example, the number 342 is 3 of 10^2 plus 4 of 10^1 plus 2 of 10^0. The base of the power 2^3 is 2 and this is the base of the binary system, with just two digits, 0 and 1

coefficient The number in front of an algebraic term. It multiplies the entire expression, for example the coefficient of $3a^2b$ is 3 and the coefficients of the terms in the expression $2a + 3bc + 5d^2$ are 2, 3 and 5. In physics some physical quantities are called coefficients – such as coefficients of friction.

component A vector can be resolved into two or more components in an infinite number of ways, just like a number can be represented by the sum of any number of numbers. Of interest to the physicist are the two components of a vector that are at right angles to each other. Given a vector of value v it may be of interest to know the component of that vector in a particular direction, say the direction of a magnetic field. If the angle between the vector and the field is θ then the component in the direction of the field is $v \cos \theta$ and the component at right angles is $v \sin \theta$.

denominator A vulgar fraction is the ratio of two numbers and is shown by placing one number above a line or a slash and the other, the denominator, below it. The denominator can be any non-zero integer or algebraic term or expression. It denominates the size of the fraction whether it be fifths or eighths for example. Once a set of fractions have been expressed with the same denominator they have the same denomination and can then be added or subtracted easily. That is why it is so useful to express fractions as percentages. The denominator is 100.

differentiating In mathematics differentiating, or differentiation, is the operation of finding the instantaneous rate of change of a function with respect to one of its variables. This is equivalent to finding the gradient of a tangent line at a point on a curve. Mathematically, if two points are placed close together on a curve and joined by a straight line, the gradient of that line is $\Delta y/\Delta x$ and the gradient of the tangent is found by making Δx approach zero. This is known as differentiating or finding the first derivative of a function. In calculus notation, $dy/dx = $ limit of $\Delta y/\Delta x$ as $\Delta x \to 0$. Some advanced level physics curricula require that students use calculus while others retain the notation $\Delta y/\Delta x$.

exponent The exponent of a number is the index of a number, or the logarithm of a number with respect to a particular base. The exponent of 5^8 is 8 and the logarithm is 8 to the base 5. If $y = 3^x$ then y is an exponential function.

g g is the symbol for the acceleration due to gravity on Earth, about 9.8 m s^{-2} depending on what part of the world you live in. Because of the Earth's rotation, oblateness and variation in height above sea level, g varies from 9.77 in Kuala Lumpur, Singapore and Mexico City to 9.83 in Oslo and Helsinki. All these values are 9.8 to two significant figures. g can also be expressed as 9.8 N kg^{-1} when referring to field strength rather than the acceleration caused by the field strength.

geometric A geometric progression is one where there is a constant ratio between consecutive terms. If the scale markers on a graph axis increased geometrically then the scale would be a logarithmic scale. 10, 100, 1000, etc. is a geometric scale of constant ratio 10 and used for many physical quantities such as loudness on the decibel scale and earthquake intensities on the Richter scale.

gradient The gradient of a slope is the vertical rise divided by the horizontal distance covered. The gradient of a line graph or tangent line in graphical analysis is $\Delta y/\Delta x$ and as this corresponds to the gradient of the tangent line it is no surprise that tan θ is opposite over adjacent in a right-angled triangle.

infinity The notion of a number which is always bigger than any number specified. Such a number is not a real number and cannot be found using arithmetic operations, therefore dividing by zero is not allowed. Richard Feynman used the concept of infinity in his "many (infinite) paths" interpretation of quantum mechanics in 1948.

integrating The inverse of differentiation is integration. It can be interpreted as finding the area under a graph. Numerical integration is the counting of squares under the graph to find the area. Integration can be applied to volume measurement by rotating functions around the x-axis. Integration may have been discovered by Archimedes, and volumes of revolution were certainly used by Kepler long before the discovery of calculus by Newton and Leibnitz. Some advanced level physics curricula use integral calculus while others require only numerical methods for integrating.

inverse Inverse variation occurs when one quantity increases while the related quantity decreases. Pressure and volume are inversely proportional to each other so that $P \propto 1/V$. The test of an inverse variation is simply to take some pairs of data and multiply them together to see if the result is constant, within experimental uncertainty. Do not confuse inverse variation with straight lines of negative gradient. A straight line of negative gradient going through the origin is still a direct proportion.

numerator The numerator is the upper part of a vulgar fraction and can be an integer or algebraic term or expression. It specifies the size of the fraction that is denominated by the denominator. 3/5 is greater than 2/5. Again the usefulness of percentages is apparent for ranking fractions.

pendulums A pendulum is any weight suspended from a pivot that can swing freely. The suspension can be rigid or flexible. If the pendulum's amplitudes are not to decrease too rapidly with time then the pivot must be as frictionless as possible. The pendulum's displacement is almost that of a simple harmonic oscillator, with constant period, which is why they were used in old clocks. A pendulum can be vertical with the weight at the top of a stiff but flexible connection to a fixed base. Such pendulums are very sensitive to variations of gravity and are used by geophysicists.

per cent Simply means "per hundred" and is a commonly accepted way of writing fractions in order to make comparisons or carry out arithmetic operations. A fraction is turned into per cent simply by multiplying by 100. In fact the multiplication is by 100/100 but the denominator is omitted and written as %.

perimeter The perimeter of a two-dimensional shape is the distance around the outside. The perimeter of a circle is given the special name circumference.

power A power is a number expressed using a base and an exponent. For example 10^4 is the fourth power of 10 where 10 is the base of the power and 4 is the index. Scientific notation uses power of 10 to express measurements. Powers of 10 are used to express orders of magnitude so that the distance to the Moon (average 384 403 km) is 10^5 km to an order of magnitude. Power is also the rate of doing work, measured in watts.

precision Measurement precision is about the closeness of agreement of repeated measurements. A measurement is precise if values cluster closely.

pyramid A cone is a pyramid with a circular base. Change the shape of the base and you can have square or hexagonal pyramids. The volume of any pyramid is $1/3 \times$ base $\times$ height, where the height is measured from the apex perpendicularly to the base.

radian A natural measure of angle. When the angle at the centre of a sector of a circle is one radian, the arc length is equal to the radius. 2π radians is therefore equal to one revolution or 360°. (Note that the gradian, which you may find on your calculator, is an angle measure found by making one right angle equal to 100 gradians.)

random errors "Errors" made by an experimenter, leading to variation in measurements. These may not be due to actual error but may arise, for example, from the lack of precision of a measuring instrument. They are more correctly termed "uncertainty".

repeatable If some degree of precision is obtained by repeating measurements with the same apparatus in the same laboratory over a short time scale, then the results are said to be repeatable. A measurement is repeatable if it gives the same or similar results under exactly the same conditions.

reproducible If some degree of precision is obtained when measurements are reproduced in another laboratory, then the results are said to be reproducible. In this situation the conditions may not be identical to the conditions when repeated. Reproducibility is important for peer confirmation of results.

resistivity Quantities can either be a property of the object or a property of the material from which the object is made. Whereas resistance, along with conductance, is a property of a particular resistor, resistivity and conductivity are properties of the material of the resistor (or conductor). Since $R = \rho L/A$, where ρ is the resistivity, ρ has units Ω m. If $A = 1$ m^2 and $L = 1$ m then the value of the resistivity is equal to the value of the resistance of a cube of the material of side 1 m, $R = \rho$.

resultant The single vector formed from adding two or more vectors together is known as the resultant. The net force on a body is the resultant force.

significant figures The number of meaningful figures to which the value of a quantity has been measured is the number of significant figures. Any further figure is beyond the precision of the measurement and is not meaningful.

simple harmonic oscillator A simple harmonic oscillator oscillates with simple harmonic motion, SHM. The period of the oscillations is constant and is independent of the amplitude of the oscillations. The acceleration of an object performing SHM is proportional to the displacement from the equilibrium position and always directed towards it: $a = -\omega^2 x$ where ω^2 is a constant equal to $4\pi^2/T^2$.

simultaneous Means at the same time. Two independent equations relating the same two variables are said to be simultaneous equations, and these allow the two unknown variables to be determined. Two simultaneous equations can be solved by substitution, by linear combination or by Cramer's rule. In general, the number of unknown variables must equal the number of independent equations if all the variables are to be determined.

systematic errors Systematic errors differ from the true value by a consistent amount, unlike random errors, which differ in an unpredictable way.

Systematic errors are true errors, in that they are mistakes due to technique or badly calibrated apparatus. For example, the zero error on an instrument is a systematic error and can be allowed for.

tare Setting a measuring instrument to zero is taring.

uniform motion Uniform motion is motion with constant or zero acceleration. SHM is not uniform motion as the acceleration varies with displacement. Uniform motion implies the existence of a constant or zero net force.

vector Quantities that require more than one (usually two) numbers for their description are vectors. Many quantities in physics require two coordinates, to describe their size and their direction, and so are vectors. Vector quantities can be added, subtracted or multiplied using vector algebra.

zero error This occurs when the value of a measured quantity is zero and the measuring instrument indicates a non-zero value. This zero error could then be subtracted from or added to all measurements. The zero error is an example of systematic error.

Appendix

DATA

Fundamental physical constants

Constant	Preferred value	Units
Avogadro constant N_A	$6.022\ 136\ 7 \times 10^{23}$	mol^{-1}
Boltzmann constant k	$1.380\ 658 \times 10^{-23}$	$J\,K^{-1}$
electron charge e	$1.602\ 177\ 33 \times 10^{-19}$	C
electron rest mass m_e	$9.109\ 389\ 7 \times 10^{-31}$	kg
Faraday constant F	$9.648\ 530\ 9 \times 10^{4}$	$C\,mol^{-1}$
gravitational constant G	$6.672\ 59 \times 10^{-11}$	$N\,m^2\,kg^{-2}$
light speed in a vacuum c	$2.997\ 924\ 58 \times 10^{8}$	$m\,s^{-1}$
molar gas constant R	$8.314\ 510$	$J\,K^{-1}\,mol^{-1}$
neutron rest mass m_n	$1.674\ 928\ 6 \times 10^{-27}$	kg
permeability of a vacuum μ_0	$4\pi \times 10^{-7}$	$H\,m^{-1}$
permittivity of a vacuum ϵ_0	$8.854\ 187\ 817 \times 10^{-12}$	$F\,m^{-1}$
Planck constant h	$6.626\ 075\ 5 \times 10^{-34}$	$J\,s$
proton rest mass m_p	$1.672\ 623\ 1 \times 10^{-27}$	kg
Stefan–Boltzmann constant σ	$5.670\ 51 \times 10^{-8}$	$W\,m^{-2}\,K^{-4}$
unified atomic mass constant u	$1.660\ 540\ 2 \times 10^{-27}$	kg

NOTE: In most calculations at advanced level, the above constants are used rounded off to two significant figures only.

a) General

electric force constant $\frac{1}{4}\pi\epsilon_0$	$9.0 \times 10^{9}\ m\,F^{-1}$
volume of 1 mole of gas at s.t.p.	$22.4 \times 10^{-2}\,m^3$
atomic mass unit u	$931.3\ MeV$
electronvolt eV	$1.60 \times 10^{-19}\ J$
Wien constant α	$2.90 \times 10^{-3}\ m\,K$

b) Earth data

Earth mass	$5.98 \times 10^{24}\ kg$
age of Earth	$\sim 4.5 \times 10^{9}\ y$
Earth radius (mean)	$6.37 \times 10^{6}\ m$
distance from Sun	$1.50 \times 10^{11}\ m$
mean gravitational field strength g	$9.81\ N\,kg^{-1}$
acceleration of free fall	$9.81\ m\,s^{-2}$
Earth constant in gravitation GM	$4.0 \times 10^{14}\ N\,m^2\,kg^{-1}$
mean atmospheric pressure	$1.01 \times 10^{5}\ Pa$
escape speed	$1.1 \times 10^{4}\ m\,s^{-1}$
solar constant	$1.37 \times 10^{3}\ W\,m^{-2}$

c) Astronomical data

Hubble constant H	70–75 km s^{-1} Mpc^{-1}
astronomical unit AU	1.5×10^{11} m
light yearly	9.46×10^{15} m $= 6.32 \times 10^4$ AU $= 0.31$ pc
parsec pc	3.09×10^{14} m $= 2.06 \times 10^5$ AU $= 3.26$ ly
age of Sun	5×10^9 y
luminosity of Sun	3.90×10^{26} W
mass of Sun	2.00×10^{30} kg
sidereal year	3.16×10^7 s
mass of Moon	7.35×10^{22} kg
radius of Moon	1.74×10^6 m
Earth–Moon distance (mean)	3.84×10^8 m
mean density of Universe	10^{-31} kg m^{-3}

Standard symbols for quantities

A displacement, amplitude, area
a acceleration
B flux density
C capacitance
c specific heat capacity
d separation
E energy, e.m.f.
F force
f frequency
I current, intensity
k any constant
I length
m, M mass
n, N number (no dimensions)
P power
p momentum, pressure
Q charge, energy transferred thermally
R resistance: electrical, thermal
r radius, separation
s distance, displacement
T Kelvin temperature
t time or Celsius temperature
U internal energy
V voltage, p.d., potential
v speed, velocity, volume
W work
x distance, displacement, extension
Δ change in quantity
θ Celsius temperature, angle
λ wavelength, decay constant
ρ density, resistivity
σ electric conductivity
ϕ flux
ω angular velocity/frequency
$\in$ electromotive force (e.m.f.)

Index